The
Electric
Power
Business

EDWIN VENNARD

President, Commonwealth Management Consultants, New York, New York

Second Edition

McGRAW-HILL BOOK COMPANY
New York   St. Louis   San Francisco   Dusseldorf
London   Mexico   Panama   Sydney   Toronto

Basic data for most of the load charts in Chapter 8 have been obtained through the courtesy of the Load Research Committee of the Association of Edison Illuminating Companies.

Preface

The electric light and power industry was founded by engineers and men trained in the physical sciences. In the early years, the major problems were technical. Solving these problems required men who knew something about the behavior of electricity. As the industry grew and came to take its place among other commercial enterprises, new problems arose. The engineers who had been perfecting the mechanics of production now became involved with problems of marketing their commodity.

These were problems of a going business—problems of economics. Because the business is one in which public interest is a strong factor, and because the electric company is a natural sole supplier in the area it serves, utility problems differ from ordinary business problems. The economic principles that apply to the operation of a corner drugstore or grocery store do not always apply to the operation of a public utility. The public utility industry has become a separate entity in the field of economics and business administration, and certain principles governing this industry have been established. Because the economic side of the business has become a specialized study, it has come to be regarded as complicated and beyond the comprehension of all except those whose jobs require them to know something about it. The practical result is that many employees do not know the basic economics of the electric utility business. They do not understand how the work they are doing fits into the whole.

Today the electric utility business is entering a new phase of development. Now practically everyone has electric service, and the future growth of the industry will, to a large extent, come from population increase and from rising use on the part of existing customers. Growing importance will be placed on economy of operation to keep the cost as low as possible, and on winning and holding public confidence. More than ever before these problems require employees who are well informed and who understand the industry.

This book describes and discusses the electric utility business as a whole. It does not go into exhaustive detail in explaining any particular phase of the business; other valuable books are available for that purpose. It is the aim of this book to present the basic facts about the power business and the economic principles that govern it. The material is presented in a simple, interesting, and understandable manner. Detail may in some cases be sacrificed for simplicity and understandability. The general principles will be laid down as accurately as limitations will permit.

The economics of the electric utility business are set against an operational background. For this reason, the engineering, accounting, financing, sales, and legal phases of the business will necessarily be dealt with. To aid in seeing the various principles in operation, a hypothetical company has been created. The components will be examined to show the relation of one to the other. The company will face much the same conditions that a real utility company faces in actual operation, so that the economic principles may be seen a little more clearly than if they were merely stated in the abstract.

The company is not patterned after any particular company. The examples will illustrate an idea or a principle of economics rather than an actual situation. Diagrams and charts are used for illustrative purposes more than to depict an actual situation. Hairline accuracy will be sacrificed if necessary to the clear grasp of a principle or a method.

The author is indebted to his associates and friends over the years who have helped in the preparation of this book. There have been so many that it would be difficult to name them without omissions.

Most of the factual information comes from the Edison Electric Institute. Where opinions are expressed, they are those of the author.

EDWIN VENNARD

Contents

1

The Growth
of an Industry

You flip a switch and the lights go on. Across town, someone presses a button and a motor starts to whir. Neither of you is excited. What has happened seems normal and not at all unusual.

For decades Americans have been taking the benefits of electricity for granted. In a sense, of course, this can be taken as the greatest possible compliment that could be given to the nation's electric utility industry. People know that when they want electricity, it will be there.

It seems hard to believe that electricity has been harnessed for only a relatively few years. Less than one hundred years ago people were still unable to make practical use of this flexible, convenient form of energy. Men and animals were doing most of the work, and fire was the direct source of light and heat.

The Beginnings of Light

It is difficult to say exactly when the electric industry was born. Most people would probably pick the day in 1879 when Thomas Edison developed his electric light, but Edison's accomplishment would not have been possible without the work of others before him. After all, people have known something about electricity for centuries.

The ancient Chinese, for example, were able to construct a magnetic compass and learned to induce magnetism in iron and steel. Centuries later, some 600 years before the birth of Christ, a Greek philosopher named Thales found he could produce static electricity by rubbing amber with a piece of cloth. Doing this, he discovered, gave amber the power to attract small bits of wood, feathers, leaves, and other light objects.

No one seems to have understood the importance of Thales' discovery. For hundreds of years the new source of energy was virtually ignored. Not until the seventeenth and eighteenth centuries did men begin to learn very much about electricity.

However, once scientists like Alessandro Volta in Italy, Benjamin Franklin and Joseph Henry in America, and André-Marie Ampère in France began to experiment with electric energy, the body of knowledge grew rapidly. By 1808 an Englishman named Sir Humphry Davy was demonstrating crude forms of both arc and incandescent lighting. The arc light consisted of a glowing spark produced by making an electric current jump the gap between two pieces of carbon. The incandescent light was made by passing a current through a piece of wire, causing a white hot glow. While the forms Davy's inventions took were not practical, they did point to the coming revolution in lighting.

A Continuous Current

In 1831 another English scientist and inventor, Michael Faraday, completed an equally important series of experiments. People had known for years that electricity somehow produced magnetism. But one day Faraday asked himself whether the reverse could also be true. Could magnetism produce electricity? After many trials, he discovered that by moving a magnet through a coil of wire he got a pulse of electricity. Doing this, he learned, made it possible to generate a continuous current of electric energy. Faraday's discovery forms the basis for most of the electric generating equipment in the world today.

By the middle of the nineteenth century electricity was being put to use in a number of practical inventions. Samuel Morse, an American painter, had developed the telegraph. Joseph Henry had constructed an electromagnetic motor. The first crude applications of electricity to transportation had been made with an electric carriage, an electric locomotive, and even an electric paddle-wheel boat.

As the century rushed on, more inventions came tumbling from the fertile minds of young inventors. In 1876 Alexander Graham Bell developed his telephone. The next year Charles Brush, a Cleveland chemist, designed a practical arc-lighting system. Two years later, in September, 1879, an electric arc-lighting central station using the Brush system, the California Electric Company of San Francisco, was opened. It was probably the first company in the world set up to produce and sell electric service. Its first customers were commercial establishments and factories. The company prospectus stated that the new arc light was "not offered for domestic purposes, because in dwellings it is not as cheap as gas or oil and is not yet adapted to such uses."

The arc light startled the country. Brush displayed his system at fairs, at scientific expositions, and in circuses. He had to let people see the brilliance of the light and show them that a new kind of artificial light was available. Generally, arc lighting was used outdoors, and soon the huge lamps, placed on high towers, began to spring up in town squares and along city streets.

One Midwestern editorial writer described his reaction to the arc light this way: "Electricity is developed by violence; that is, by waste and disturbance of atoms of matter, which is necessarily expensive. For sensational uses, for spectacles where expense is of minor consideration, electric light will, of course, be employed. But the great mass of the people will never be able to use this costly illuminator to banish darkness from their humble dwellings."

Despite such pessimism, other inventors rushed into the field. In Europe and America, competing arc-lighting systems were developed. Swan, Siemens, Jablochkov, Fitzgerald, Weston-Maxim, and Thomson-Houston were a few of the leaders. But as their systems were being installed in London, Paris, and various American cities, a powerful competitor was being born that, eventually, would force them all to the side.

Edison Enters the Field

Thomas Edison turned his attention to the problems of electric lighting in 1877. Although he was just thirty years old, he had already won a worldwide reputation for his improvements in the telegraph and telephone. Even then, newspapers were calling him the "Wizard of Menlo Park."

Characteristically, Edison approached the new field with a breadth of vision and hard practical sense unequaled by his rival inventors. From the first, he turned his back on the arc light. He had made up his mind "to replace lighting by gas by lighting by electricity. To improve illumination to such an extent as to meet all requirements of natural, artificial, and commercial conditions . . . not to make a large or a blinding light, but a small light having the mildness of gas."

To accomplish his purpose, Edison had to do six things. He had to develop an economical generator that would manufacture a steady flow of electric current. He had to devise a system of conductors that could be tapped at frequent intervals so that electricity could be piped into individual homes. He had to work out a method of maintaining constant voltage through the system, no matter how many lamps were used. He had to invent a lamp that would give better, less expensive light than the gas jets then in use. He had to develop a way of feeding current into a multiple circuit that would permit operation of any number of light bulbs, and he had to invent a meter that would measure and record the amount of electricity being supplied to each customer.

Edison knew that about 90 percent of the revenues of American gas companies came from residential and office lighting and that the arc light was unsuitable for these uses. He also realized that a practical electric lighting system would be a great boon to the people of the world, and that the idea of electric lighting would capture the imagination of the public. But, in his notebooks he wrote: "It doesn't matter if electricity is used for lighting or for power. Small motors can be used night and day." And later he noted: "Generally, poorest district for light, best for power, thus evening up the whole city—note the effect of this on investment." From the beginning, he seems to have understood that lighting alone would not be enough to support an electric generating company.

Developing a complete electric system, Edison realized, was going to be a long, hard job. It would take machinery, parts, and tools. It would take the help of a larger staff than he had ever used. Above all, it would take money.

By careful use of publicity the inventor was able to maintain the public's interest during the years he worked on the problems facing him. His brash newspaper statements and his exhibits of progress at Menlo Park served another purpose as well: He kept investors aware of his activities and ready to supply him with the money needed to carry on experiments. As it turned out, more than $235,000 was spent on experimental work, patent expenses, and general overhead between October, 1878, and September, 1881, alone. More than $400,000 more was needed during the following five years.

By forming the Edison Electric Light Company and selling shares of stock, Edison was able to attract a group of men into backing his experiments. As stockholders in a company, each man could risk as much money as he could afford, and together the group was able to provide the funds Edison needed. The twelve men who were the initial investors in the company must have been drawn principally by the inventor's bold vision of a commercial electric light and power system. But his experience in telegraphy must have had an important bearing on their decision to join in the venture, too. After all, eight of the twelve were directly involved in the telegraph industry themselves.

Working steadily, and during certain periods almost night and day, Edison and his staff accomplished all they set out to do in an incredibly short time. On October 21, 1879, they put an incandescent electric light bulb in circuit and kept it lighted for forty hours. By 1881 all the essential components for a commercially successful central-station lighting system had been developed.

Two Kinds of Customers

Eventually, Edison and his backers hoped to be able to sell the new lighting system to two kinds of customers. First, there would be the people who wanted to produce light for their own use. These would include factories, hotels, large stores, and private mansions. Second, the system would be sold to people who wanted to produce and sell electricity to others. Before this could be done, however, Edison would have to build, test, and operate a central generating station himself.

As early as the spring of 1880, only three years after he had started to work on the problem of electric lighting, Edison installed his first "isolated plant" on the steamship "Columbia," which sailed around Cape Horn to California with its 115 lamps blazing almost continuously. By November, 1881, there were seven more isolated plants in operation, and the Edison Company for Isolated Lighting had been set up and licensed to meet the expected demand. Within the next year the company installed 145 more isolated plants.

Edison's main attention, however, was directed to the construction of his central generating station. Located on Pearl Street in downtown New York City, the station was small and, by modern standards, relatively primitive. It could transmit energy only a little over 5,000 feet—less than a mile. It had six Jumbo generators, which had capacities of 120 kilowatts each. And when it began operation on September 4, 1882, it had a total of fifty-nine customers.

The Pearl Street station was operated by steam. Later the same month the second Edison station, operated by water power, opened at Appleton, Wisconsin. Thus, what are still the two chief methods of producing electricity have been used since the industry's earliest days.

Edison's system was the first to make house-to-house distribution of electric power practicable. But it had serious drawbacks. For example, the distribution system required such an expensive investment in copper cables that the area a generating station could serve was severely limited. The stations had to be located in the immediate area where the electricity was to be used. Better, more economic methods of producing and transmitting energy had to be developed before electricity could be brought to any but the largest, most densely populated cities.

In a relatively short time, however, a series of inventions and improvements made it possible for costs to be lowered so that somewhat less densely populated areas could be served. A distribution system was developed that radically reduced the amount of copper needed and thus substantially lowered the costs of building distribution lines. The reduction of energy consumed by incandescent lamps from 6.5 watts per candle in 1882 to 3.1 watts in 1890 made the use of electricity more economical for consumers.

Possibly most important, the development of the alternating current system by George Westinghouse and his chief engineer, William Stanley, made long-distance transmission possible. Stanley's transformer, first demonstrated in 1886, along with his alternating machine and his method of connecting transformers in parallel, allowed electric power to be generated at a low voltage, stepped up to a higher voltage for more economical transmission, and then stepped back down to a lower voltage for use. The result was that central stations no longer had to be located in the areas where electricity was being used. They could be near fuel facilities and water in low-rent areas, and the stations could be larger, permitting further economies.

John W. Lieb, who was the first electrician of the Pearl Street station, recalled the first years of the industry in these words: "Without a clear idea of what was required and without any engineering precedents to follow, central station pioneering was largely a groping in the dark, an endeavor to meet intuitively or by unlimited expenditure of personal energy and resourcefulness, the unexpected problems which daily presented themselves, and which often needed instant solution."

"It is remarkable," William Stanley said some years later, "how little of modern electricity we all knew in those days. The terms inductance, self-induction, armature reaction, transformer, converter, single phase, multiphase, hysteresis, and a number more were unknown. When we saw phenomena that we did not understand we had to try to fit some sort of names to them. It was some years later before the art was ready for high potential work, for it was necessary that the alternating machines, the transformers, and the system devised should attain a certain degree of perfection before engineers dared to introduce the idea of serving electricity over great distances by means of very high potentials, as the profession understood that term."

Something More than Light

In the early days of the electric industry, customers of the central stations looked on electricity as a luxury. They had good reason. The costs of setting up a central station, with its expensive generating equipment and distribution systems, were extremely high. The bulk of electricity being sold was used for lighting streets and commercial establishments, and in 1896 more than half of the central stations in existence only operated in the hours after dark.

It was not until the 1890s that central station managers began to understand that if electricity was to be seen as a necessity rather than a luxury, the rates to consumers would have to be rearranged. A method had to be worked out so that the customer who made use of the company's investment most steadily during the year would be charged the lowest possible price.

Equally important, other uses for electricity had to be developed. For all of recorded history, men have known that they could multiply the amount of work they could do by using tools. Before the harnessing of electricity, there had to be the closest possible link between the tool and its source of energy. This was true whether the tool was an ax and the energy came from a man's arm, or whether the tool was a water wheel and the energy came from a rushing stream. Systems of gears, pulleys, and belts were devised, but the distance between the tool and the source of power was still limited. The coming of the electric motor changed all that.

The electric motor made it possible for a tool to be far removed from the source of power. Electricity provided a clean, quiet form of power that could be sent by wire directly to the spot where it was needed.

By 1893 an efficient generator had been developed, an economic method of transferring energy from the generating point to the point of use had been worked out, and reliable motors were being manufactured. That year, one electrical manufacturer installed a dozen 65-horsepower alternating current motors in the textile mills at Columbia, South Carolina. The door had opened to an enormous potential market.

The electric motor was basic to the success of the infant electric companies. Development of uses for the motors made it possible for the companies to expand from the part-time operation of producing light to the full-time operation of producing light and power.

The First Electric Systems

After 1900 one of the most notable tendencies in the industry was toward the consolidation of small, individually owned stations into larger systems. This tendency, made possible by increasing knowledge of higher voltage transmission combined with extensive installation of generating and consuming equipment of higher and higher efficiency, brought cheaper and better service to the public.

People had discovered that they could get the best electric service at the lowest possible price if a single company served as the sole supplier in a given area. Joining small companies together into efficient systems made it possible to lower costs by doing away with unnecessary duplications of equipment. With the cost of production down, the price could come down too.

But consolidation of this kind took capital. In the beginning, the typical central station was owned by one man. Corporate organization, with its greater facility for attracting investment and financing, came slowly. By 1902, however, 73 percent of the 2,805 investor-owned central stations were owned by companies, and the percentage increased year by year.

The Pattern of Growth

By 1910 electric utility men were able to see a pattern in the growth of their industry. Like any new business, the companies had first served the market that was immediately available and promised the surest return: the thickly populated centers. Experience in producing and selling electricity, coupled with technical advances, then made it possible for the companies to serve smaller population centers. With further technical advances it became possible to bring electricity to even smaller towns and villages. In this way, electric service moved step by step from the large cities, to the towns, to the small villages and rural areas, until today, only about ninety years after Edison developed his light bulb, electric service is available to virtually everyone in the country.

The pattern of growth also has been characterized (1) by development and use of larger and larger generating units using steam at higher and higher temperatures and pressures and (2) by increasing development and use of interconnections, using transmission lines of higher and higher voltages. Both of these have been important to the development of today's highly reliable electric service and to improvements in the efficiency in converting raw fuels (coal, gas, oil, falling water, and nuclear) to electric energy.

The Electric Utility Industry Today

At present, after years of effort in advancing the process of interconnection, the Electric Utility Industry is grouped into about twelve interconnected and coordinated areas. Within these areas smaller groups of suppliers of electric energy have pooled their resources for construction and operation of generating plants to provide the greatest possible economy of operation.

As the use of electricity in the country doubles about every ten years, power suppliers must double their capacity at about the same rate. Almost every power plant built is a little larger and a little more efficient than the previous plants. Advances are made in knowledge of techniques to control very high voltages and in pooling. This reaching out for improvement presents new problems which, in turn, require innovation, research, study, experimentation, and experience.

The process of evolution and growth promises to continue into the future. Since 1954, a wholly new concept of fuel has been developed. Nuclear power has come of age. The process required enormous research and development. Now on the horizon is a nuclear reactor which will breed more fuel than it uses, giving promise of further fuel economy and more economical production of electricity.

By the end of 1968 the electric utility industry in the United States had about 70 million customers—some residential, some farm, some commercial, and some industrial. Of these, almost 80 percent were served by investor-owned electric companies, about 8 percent were served by rural electric cooperatives, and the remainder were served by Federal agencies, municipal governments, power districts, and state projects.

The owners of the electric companies come from every walk of American life. In fact, almost everyone in the country has at least an indirect financial interest in electric utility company operation. Some 4 million stockholders are the direct owners of the electric companies, but in addition, millions of life insurance policyholders and 23 million mutual savings bank depositors are among the indirect owners. Pension funds, trusts, charitable and fraternal societies, religious and educational institutions, and foundations also have interests in electric companies.

More than 350,000 people are employed in electric utility

companies in the United States. Adding the employees of the manufacturers of electric equipment, apparatus, and appliances, some 3 million people are employed in the entire electrical industry. Millions more are employed in other electronics, radio and television broadcasting, motion picture, light metals, and chemical industries, none of which would be possible without abundant electricity.

The harnessing of electricity has resulted in millions of new jobs, the creation of entire new industries, and great advances in science and technology. But the best is yet to come. As Thomas Edison put it in 1928, "So long as there remains a single task being done by men or women which electricity could do as well, so long will that development be incomplete. What this development will mean in comfort, leisure, and in the opportunity for the larger life of the spirit, we have only begun to realize."

2

Trade Terms

The Kilowatt and the Kilowatt-hour

Like any other industry, the electric power industry has its trade terms. The people who work in the business have developed a vocabulary peculiar to this industry. Understanding these terms is one of the keys to understanding the electric power business.

One of the distinguishing characteristics of electricity is that it cannot be stored economically in large quantities. It must be generated as it is used. It is for this reason that it is important to distinguish between the quantity of electricity used and generated and the *rate* at which it is used and generated.

The *kilowatt-hour,* for example, is the measure of the quantity of electricity generated or consumed. The kilo-

MEASURES OF QUANTITY

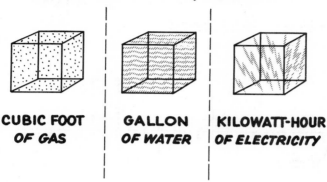

CUBIC FOOT	GALLON	KILOWATT-HOUR
OF GAS	OF WATER	OF ELECTRICITY

CHART 2.1

watt-hour corresponds to other units of measure of quantity, as shown in Chart 2.1. A cubic foot of gas, a gallon of water, and a kilowatt-hour are all measures of quantity. A dozen oranges, a pound of bread, and a box of pins are all similar measures. They tell how much of the item there is. For example, "10 gallons of water" conveys an idea of the amount of water. It does not tell the listener whether the water is being used, or whether it is flowing through a pipe; it merely identifies a quantity. The kilowatt-hour tells the same thing about electricity. It is simply a measure of quantity.

Water Analogy

When talking about *how fast* a person is using water, both the quantity—gallons—and the time during which the quantity is used become important. For example, Chart 2.2 shows how a person might use water during the day. This person had all his faucets turned off during the night, as most people do, and used no water until 7 o'clock in the morning. Between 7 and 8 A.M. he used 1 gallon, and the chart accordingly shows a cube representing a gallon of water used during that time.

USE of WATER
(EXAMPLE)

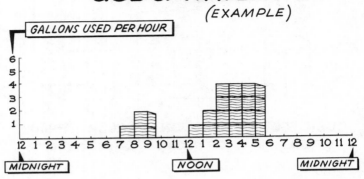

CHART 2.2

Between 8 and 9 o'clock, he used 2 gallons of water, and two cubes are shown in the space representing that hour. Then from 9 until noon he used no water. From noon until 1 o'clock he consumed a gallon of water; between 1 and 2 o'clock he used 2 gallons. Then at 2 o'clock, perhaps he turned on his lawn sprinkler, which used 4 gallons between the hours of 2 and 3. He let the sprinkler run until 5 o'clock, and it continued to use 4 gallons in each hour until that time. At 5 o'clock, he shut off the sprinkler, and used no more water that day.

Counting the number of cubes on the chart shows that he used a total of 18 gallons of water that day. The chart also shows how fast he used water. From 7 to 8 o'clock in the morning, he was using water at the rate of 1 gallon per hour. During the next hour he used water at the rate of 2 gallons per hour. His maximum rate of use occurred when he was running his sprinkler, and during that time he used water at the rate of 4 gallons per hour.

Electricity Example

In the water system, water flows through a pipe. In the electric system, energy, in the form of electricity, in effect

flows through a wire.   The energy is used in producing light
and heat, and to run motors.

The quantity and rate of use of electricity are measured as
in the water example, the only difference being in the terms
used to describe the units.   The unit of measurement of
quantity of electricity is the kilowatt-hour.   The abbreviation
of the kilowatt-hour is kwhr.

Take a look at this same person's use of electricity during
the day.   Chart 2.3 shows how it might look.   In this chart,
the cubes represent kilowatt-hours of electricity.

Here again he is using no electricity during the night.
When he gets up in the morning, his wife uses the electric
range to cook breakfast, and may plug in the toaster.   This
takes 1 kilowatt-hour between 8 and 9 o'clock, and a cube is
shown for that hour.   He makes no further use of electricity
until late in the morning, around 11 o'clock, when his wife
turns on the washing machine to do the laundry.   She also
uses the dryer for a short while and uses 3 kilowatt-hours between 11 and 12 o'clock. At noon, in cooking lunch and perhaps baking a cake, she uses 4 kilowatt-hours.   Then there is
no further use until evening, when she begins to prepare din-
ner.   Between 5 and 6, she uses 3 kilowatt-hours, and be-

CHART 2.3

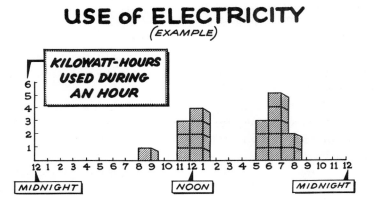

USE of ELECTRICITY
(EXAMPLE)

tween 6 and 7 she uses 5 kilowatt-hours. It is possible that during this hour she is doing considerable cooking, and also since dusk is approaching, she has the lights turned on. Perhaps the children are using electricity to run the television set.

After the cooking is finished, her use of electricity drops so that between 7 and 8 P.M., only 2 kilowatt-hours are used. At 8 o'clock, the whole family may leave to spend the night at Aunt Minnie's, and no electricity is used for the remainder of the day.

Counting the cubes tells how many kilowatt-hours this family used during the day. The total is 18. To find out *how fast* this family uses electricity, the number of kilowatt-hours used during any hour can be tallied to find the rate of use. From 8 to 9 in the morning, for example, 1 kilowatt-hour was used. The highest rate of use occurred between 6 and 7 o'clock; 5 kilowatt-hours were used during this hour.

Power means the rate at which energy is delivered or used and is analogous to the term "gallons per minute," or "gallons per hour." The power of an automobile engine is rated in *horsepower,* abbreviated hp. This rating indicates the engine's capacity to deliver the energy required to run the automobile.

The *kilowatt* is the same kind of unit as the horsepower. It measures the rate at which electricity is being generated or consumed. The abbreviation of kilowatt is kw.

Electric appliances are rated in watts and kilowatts. The *watt* is $\frac{1}{1,000}$ of a kilowatt; in other words, the kilowatt is equal to 1,000 watts. The watt is used to rate appliances using small amounts of electricity.

The *ampere* is sometimes used to indicate the rating of appliances, fuses, wires, etc., in the home. The ampere is the measure of the flow of current, as distinguished from voltage, which is pressure.

If an appliance is rated at 6 amperes and the house voltage is 115 volts, the watts are 690 (115×6). At unity power factor the watt may be referred to as a *volt-ampere*. Also a kilo-watt is a *kilovolt-ampere* (kva), a kilovolt (kv) being 1,000 volts. (See page 28 for a discussion of power factor.)

A light bulb may be rated at 100 watts, which is equal to 1/10 of a kilowatt. This rating means that the lamp will use 1/10 of a kilowatt-hour of electricity during each hour it is used. If such a lamp is burned steadily for 60 hours, it will have used 6 kilowatt-hours (1/10 × 60).

Chart 2.4 shows use of electricity again, but this time uses the term kilowatt to describe rate of use. Note that now the vertical scale reads in kilowatts.

During the hour between 6 and 7 o'clock, this family used 5 kilowatt-hours of electricity. During that hour, they used electricity at the rate of 5 kilowatts (meaning at the rate of 5 kilowatt-hours per hour).

It is important to the power company to know how much electricity people are likely to use during any period, so that the company can have enough generators to make that electricity when it is needed. The company's generators are rated according to how fast they can turn out units of electricity. A generator may have a rating of 500,000 kilowatts, which means that this machine if run for an hour can make

CHART 2.4

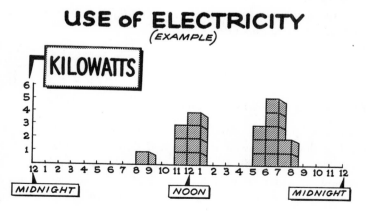

USE of ELECTRICITY
(EXAMPLE)

500,000 kilowatt-hours of electricity. If it ran for two hours, it would make 1,000,000 kilowatt-hours. If it ran for half an hour, it would make 250,000 kilowatt-hours. Also the power company must have neighborhood transformers and power lines big enough to take care of the customer's highest rate of use of electricity.

The power company which serves the family in the example has to have enough generating capacity to take care of that family's needs and the needs of all the other customers it serves. Referring again to Chart 2.4, it is noted that between 6 and 7 o'clock at night this family used 5 kilowatt-hours. That means that the power company must be able to send out 5 kilowatt-hours during that hour for this customer.

The rating of an appliance in watts and kilowatts enables a determination of how many kilowatt-hours of electricity the appliance is using. To find out *how much* electricity is used the length of time the bulb or appliance is in use at that rate must be known. If ten 100-watt bulbs using electricity at the rate of 1 kilowatt are burned for one hour, they will use 1 kilowatt-hour of electricity. If twenty such bulbs are burned for an hour, they will use 2 kilowatt-hours.

In summary, then, the *kilowatt-hour* is the basic measuring unit for telling *how much* electricity has been delivered or used; the kilowatt tells *how fast* these units were used.

Table 2.1 shows how these terms compare with other measures of quantity and rate.

Chart 2.5 shows some examples of measurement of rate of use.

Electric System—Water Analogy

Because people cannot see, smell, or hear electricity, it is hard to describe how it is made and delivered without using an analogy. Since the flow of electricity through a wire is like the flow of water through a pipe, electrical principles can be shown using water as an illustration. Some of the terms used

TABLE 2.1   Terms of Quantity and Rate

Item	Quantity	Rate
Water.	Gallons	Gallons per hour
Travel.	Miles	Miles per hour
Oil production.	Barrels	Barrels per day
Printing press	Impressions	Impressions per hour
Gas	Cubic feet	Cubic feet per hour
Factory machine.	Units	Units per hour
Milk.	Gallons	Gallons per day
Electricity	Kilowatt-hours	Kilowatts °

° Note that the term kilowatt includes a time element. It could also be stated as "kilowatt-hours per hour," telling how fast these units were made or used.

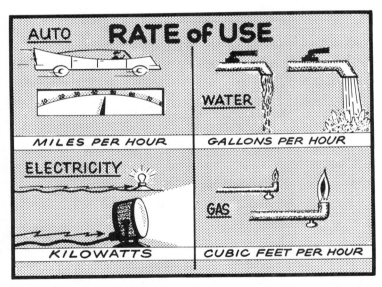

CHART 2.5

to describe the water system are also used in the electric business.

Chart 2.6 shows a water system serving three customers. (Some water systems have water-storage facilities to meet the customer's demands when the pump is not running. Other

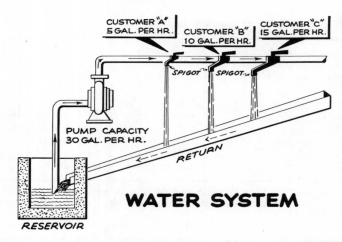

WATER SYSTEM

CHART 2.6

systems have no storage facilities and the pump must be running at all times. A system with no storage facilities has been used in this illustration to compare with electric systems, because storage is not practical in electric systems.)

The water pump, with the *capacity* to pump water at the rate of 30 gallons per hour, takes the water from the reservoir and delivers it to the three customers as they want to use it.

When any one of the customers turns his spigot, he will get water at the *pressure* that is maintained in the pipe. The pump must be kept running at all times to keep up that pressure so it will be there whenever the customers choose to use the water. The power needed to run the pump when no water is being used is merely enough to overcome the loss caused by leakage and friction in the pump.

If customer A wants water, he turns on his spigot, which takes water at the rate of 5 gallons per hour. A *current* of water is said to be *flowing* through the pipe. Immediately, the pump delivers water at the rate of 5 gallons per hour.

Assume now that customer B turns on his water at the same time A is taking water. B's spigot takes 10 gallons per hour, so that the two of them together use 15 gallons per hour and the pump delivers water at this rate. It may be said that customer B makes a *demand* for water on the water system at the rate of 10 gallons per hour. Customer A demands water at 5 gallons per hour. The combined demand is 15 gallons per hour. There is then said to be a *load* on the water pump of 15 gallons per hour. The terms "load" and "demand" are synonymous.

Now customer C begins using water at the same time. His larger spigot uses water at the rate of 15 gallons per hour. With all three customers using water at the same time, the load on the pump or plant is 30 gallons per hour, equal to the entire capacity of the pump.

The pump and pipes must be big enough (i.e., have sufficient capacity) to meet the maximum combined requirements of all customers at any one time.

The Electric System

Chart 2.7 shows an electric system similar to the water system in Chart 2.6. Fuel for the system is taken from the fuel reservoir, just as water was taken from the water reservoir. The fuel may be coal, oil, gas, lignite, or nuclear fuel. (Elec-

CHART 2.7

ELECTRIC SYSTEM

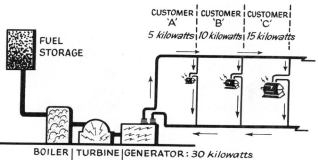

CUSTOMER | CUSTOMER | CUSTOMER
'A' | 'B' | 'C'
5 kilowatts | *10 kilowatts* | *15 kilowatts*

FUEL STORAGE

BOILER | TURBINE | GENERATOR: *30 kilowatts*
meaning ... 30 kilowatt-hours can be generated in an hour.

tricity is also generated using the energy in falling water. This is discussed in Chapter 11, beginning on page 278. The relative economy of the two processes is compared there.)

The electric generating plant converts the energy in the fuel to *electric energy*. The fuel is burned under a boiler where the energy in the fuel is converted to heat and pressure in the steam. The steam pressure spins a steam turbine which is directly connected to the electric generator, where the energy of the turbine is converted to electric energy.

If the customers are not using any electricity, the generator running at full speed merely keeps up the level of electric pressure on the system. The generator uses only enough fuel in this case to overcome the losses caused by friction in the turbine and the generator and the losses in the system.

It may seem strange that a generator running at full speed may require only a small amount of fuel. The situation is, in fact, something like that of a man whose job is to push a wheelbarrow (Chart 2.8). When there is no load in the wheelbarrow, it is very easy to push. When the wheelbarrow is heavily loaded, it may take a great deal more energy to move it. Although the load of electricity on a generator is invisible, it is there, and it takes more energy to turn a loaded generator than one that is not loaded.

CHART 2.8

ENERGY AND LOAD

In the diagram of a simple electrical system shown in Chart 2.7, the electric generator, or power plant, has a capacity of 30 kilowatts. It can deliver electric energy at the rate of 30 kilowatts (at 30 kilowatt-hours every hour) during every hour it is operated. This generator is kept running day and night, ready to supply any and all demands for electricity as they are made by the customers. The term *demand* refers to the rate at which a customer takes energy from the electric system. A customer who is using machinery that takes 100 kilowatts to operate makes a demand on the system of 100 kilowatts. He will use 100 kilowatt-hours of electricity every hour he has the machinery running. The *maximum demand* of a customer is the highest rate at which he takes energy. The customer's demand, or rate of use of energy (kilowatts), can be measured, as well as the quantity of energy (kilowatt-hours) used during the month. The maximum demand on an electric system is the highest rate at which all customers combined take energy. Demands are measured in kilowatts.

If customer A in Chart 2.7 decides to take electricity, he throws a switch or presses a button which starts him using electricity at the rate of 5 kilowatts. He is said to be creating a demand of 5 kilowatts on the electric system, just as the customer in the water system created a demand for water by opening his spigot.

When customer B decides to operate, he takes electricity at the rate of 10 kilowatts and is said to have a demand of 10 kilowatts. If customer A and customer B both use electricity at the same time, the combined demand on the electric system is 15 kilowatts.

If, while customers A and B are operating, customer C decides to operate, the total demand on the plant rises to 30 kilowatts—its capacity. That plant cannot deliver energy at a faster rate than 30 kilowatt-hours an hour.

Electricity is generated as it is used, just as in the water analogy water is pumped only as it is used. The rate at which water is pumped is equal to the combined rate at which all the customers are taking water at any time. The rate at which generators make electricity is the sum of the combined rates at which all customers are taking electricity at the particular time.

The electric utility company must foresee the highest demands its customers will make. Then the company will plan to have enough plants and equipment to meet these requirements at all times. The demand on an electric system is sometimes called the *load* on that system.

The maximum load on an entire electric system during any particular period is sometimes called the *system peak*. For example, if the maximum hourly demand on the system is 600,000 kilowatts for a particular month, the system peak for that month is said to be 600,000 kilowatts.

In a water system, there must be enough water pumps of large enough size to meet the maximum demand of all water customers. The greater the demand in gallons per hour, the larger the pumps must be. The cost of the pumps varies with the size: The larger the capacity, the lower the investment per unit of capacity (Chart 2.9).

Similarly, in an electric system, there must be enough generators, and the generators must be large enough to meet the maximum hourly demand in kilowatts of all electric custom-

CHART 2.9

WATER PUMPS

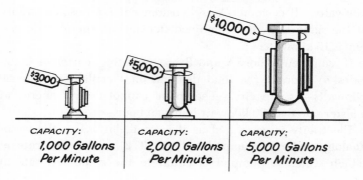

CAPACITY:
1,000 Gallons Per Minute

CAPACITY:
2,000 Gallons Per Minute

CAPACITY:
5,000 Gallons Per Minute

ELECTRIC POWER PLANTS

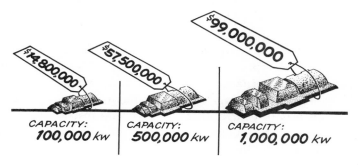

CAPACITY:
100,000 kw

CAPACITY:
500,000 kw

CAPACITY:
1,000,000 kw

CHART 2.10

ers; the greater the demand, the larger the generator—and the more expensive (Chart 2.10).

The *voltage* of an electric system is the measure of electric pressure and is analogous to water pressure in a water system. For a pipe of a given size, raising the water pressure increases the capacity of the pipe to deliver water in gallons per hour (Chart 2.11).

CHART 2.11

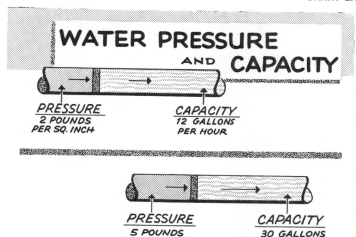

WATER PRESSURE
AND CAPACITY

PRESSURE
2 POUNDS
PER SQ. INCH

CAPACITY
12 GALLONS
PER HOUR

PRESSURE
5 POUNDS
PER SQ. INCH

CAPACITY
30 GALLONS
PER HOUR

Similarly, for a wire of a given size, raising the voltage (pressure) results in an increase in the capacity of that wire to deliver energy. This capacity is measured in kilowatts (Chart 2.12).

High voltage is used in transmitting power for the same reason that high pressure is used to transmit water or oil in pipes over long distances; smaller, and therefore less costly, wire can be used to carry a given amount of electricity. Bigger wires are, of course, heavier, which means that heavier, more expensive transmission towers would have to be built to carry the weight of the wires.

Power Factor

Certain electric devices have a peculiar characteristic which results in a demand for more kilowatts than are actually put to any useful purpose. The *induction motor*—the type of motor in most common use—has this characteristic,

CHART 2.12

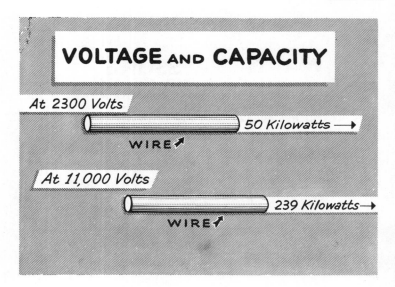

VOLTAGE AND CAPACITY

At 2300 Volts

WIRE — 50 Kilowatts →

At 11,000 Volts

WIRE — 239 Kilowatts →

when the motor is run at less than its full load. The actual work being done by the motor results in a certain kilowatt demand that can be measured by an ordinary demand meter. However, when partially loaded, the motor makes a different and useless kind of demand, above the partial load on it, on the electric system, which cannot be measured by the ordinary meter. This other demand requires capacity in the electric system just the same as the useful demand.

In other words, in the operation of such a motor there is a measurable amount of useful power and also a certain amount of useless electric current. This useless current requires capacity in the system, and there must be investment to provide that capacity. Thus, cutting down the useless power helps cut down the cost of providing electric service.

Power factor is an expression of the relationship between the useful current and the total current used in an electrical device.

The watt is the basic unit of power. There is a mathematical relationship between the watt, the volt, and the ampere, as expressed by the familiar formula:

$$\text{Watts} = \text{volts} \times \text{amperes}$$

or $\qquad$ $$\text{Kilowatts} = \text{kilovolts} \times \text{amperes}$$

The term on the right-hand side is called the kilovolt-ampere. This equation is a correct one only where no useless current is in evidence. There are certain types of electrical devices that do not cause any of this useless current in the electric system. Resistance heating devices, such as the electric range, hot plate, toaster, and incandescent electric light (which is a heating device burning white hot), cause no useless current in the system. Gaseous tube lights, such as fluorescent and neon lights, cause some useless current which can and should be corrected at the light source. When there is no such useless current in evidence, the power factor is said to be 100 percent.

The above formula, therefore, is correct at a power factor of 100 percent. The formula may be expressed as follows:

$$Kilowatts = kilovolt\text{-}amperes \times power\ factor$$

In the case of induction motors the power factor may be as low as 50 percent and is sometimes less. A 50 percent power factor means that the useful power is only half that needed to run the motor. The other half of the required power is useless. If an electric motor needs 100 kilowatts of useful power and is operating at 50 percent power factor, it would require 200 kilovolt-amperes of capacity in the electric system. The electric company must provide 200 kilovolt-amperes of capacity in transformers, lines, and generators to serve that motor. If the power factor at that motor could be raised to 100 percent, the motor would not do any more work than it did before, but it would make a demand of only 100 kilowatts, and only 100 kilowatts of capacity would be needed to serve the motor.

Fortunately, there are devices called *condensers* to correct poor power factor. There are *static condensers* (those that have no moving parts) and *synchronous condensers* (those that have rotating parts). A natural place to correct the power factor is at the device causing the poor power factor. A customer that has equipment with poor power factor may correct it by installing a condenser. By means of this device, the power factor can be raised to any desired point.

Electric utility companies usually install condensers in their electric systems to correct system power factor. These are synchronous condensers or capacitors.

Load Factor

Each customer at some time during any given month makes his maximum demand on the electric system. This happens when he is using a great number of his appliances at the same time, perhaps, and thus uses electricity at his most rapid rate. He very rarely keeps up that high rate of demand continuously; for example, a customer having a maximum demand of 5 kilowatts at some time during the month does not have a demand of 5 kilowatts during every hour of the month. During some of the hours his demand may be only 1 kilowatt. At other times it may be only half a kilowatt, and in some hours at night he may have no demand at all.

As a result, his use in kilowatt-hours is less than it would be if he used electricity at the level of his maximum demand all the time. The relationship between what he actually uses and what he would use if he used electricity continuously at the rate of his maximum demand is referred to as *load factor*.

In Chart 2.13 each square represents 1 kilowatt-hour. The chart shows how a customer who runs a small business might use electricity during a day. Between midnight and 6 A.M. he is using no kilowatt-hours. Between 6 and 7 A.M. he uses 1 kilowatt-hour. Between 7 and 8 A.M. he uses 2. Between the hours of 11 and 12 he uses 6 kilowatt-hours. Between 12 and 1 o'clock his people go to lunch and his use is only 2 kilo-

CHART 2.13

DAILY LOAD CURVE
SMALL LIGHT AND POWER CUSTOMER

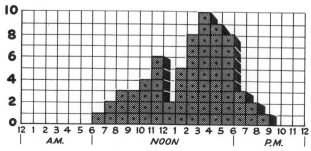

watt-hours.   After lunch he resumes his operations and reaches a peak demand or a peak use of electricity during that day between 3 and 4 P.M. when he uses 10 kilowatt-hours.   His use then falls off until he is using only 1 kilowatt-hour between 8 and 9 P.M. and no kilowatt-hours from 9 P.M. to midnight.

There is a dot in each of the squares where he was using electricity.   If all the dots are counted, it is found that he used 67 kilowatt-hours during that day.   His maximum rate of use was 10 kilowatt-hours during the 1 hour between 3 and 4 P.M.   He had a maximum demand of 10 kilowatts.

Every square under the 10-kilowatt mark represents a kilowatt-hour which the customer could have used if he had taken energy uniformly during the day.   If all of these squares are counted it will be found that there are 240.   If the customer had used energy uniformly at the rate of 10 kilowatts during every hour of the day, he would have used 240 kilowatt-hours and he would be said to have a load factor of 100 percent.   The customer actually used 67 kilowatt-hours. The load factor is the ratio of the actual kilowatt-hours used to the kilowatt-hours that would have been used if the customer had used energy uniformly during the day at the rate of maximum use.   Therefore this customer's load factor for that day was 67 divided by 240, or 28 percent.

A three-dimensional view of a customer's use for a one-month period is illustrated in Chart 2.14.

In this case the customer's use of electricity follows about the same pattern each day with a slightly different pattern for Saturday and Sunday.   From a low level during the early morning hours, his use rises to a morning peak of about 6 kilowatts.   He curtails his use somewhat during the noon hour, raises it to an afternoon peak of 10 kilowatts, and drops it to only 1 kilowatt after the evening activity.   On Saturdays he uses less and on Sunday he uses almost none.

This customer makes a peak demand on the power system of 10 kilowatts.   This means that the power company must

MONTHLY LOAD
SMALL LIGHT AND POWER CUSTOMER

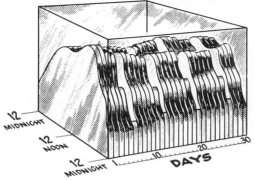

CHART 2.14

reserve that amount of capacity for the customer's use when he demands it regardless of the time of day or year. The company is standing by every hour ready to supply electricity at this rate, even though that demand may be used only a few hours during the month. The customer actually used 1,600 kilowatt-hours during the month. He might have used 7,300 kilowatt-hours if he had used the service uniformly during the month at the rate of 10 kilowatt-hours per hour or 10 kilowatts. The average number of hours in a month $\frac{365 \times 24}{12} = 730$. This customer is said to have a load factor of 22 percent for the month. This is obtained by dividing the 1,600 kilowatt-hours actually used by the 7,300 kilowatt-hours that might have been used, if the customer had operated uniformly at the peak rate of use.

Customer load factors may vary from as low as 0 to almost 100 percent, depending upon the kind of business in which the power is used.

In a similar fashion the load factor for the whole company can be determined (Chart 2.15). The load factor on a power

A COMPANY MONTHLY LOAD CURVE

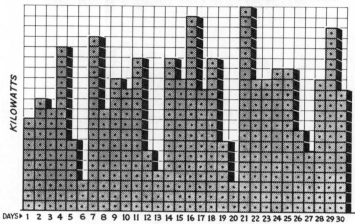

KILOWATTS

DAYS▶ 1 2 3 4 5 6 7 8 9 10 11 12 13 14 15 16 17 18 19 20 21 22 23 24 25 26 27 28 29 30

Load Factor can be found by dividing the number of shaded squares by the total number of squares...

CHART 2.15

system for any month is the ratio of the kilowatt-hours actually delivered to the number of kilowatt-hours that could have been delivered if the maximum demand had prevailed every hour of every day in the month. For example, assume an electric system delivered 500 million kilowatt-hours during a certain month and that the maximum rate of delivery at any one time was 1.2 million kilowatts (this being the maximum demand on the system). If that system had delivered 1.2 million kilowatt-hours every hour for the whole 730 hours in the month, it would have delivered 876 million kilowatt-hours. The load factor for the month was 500 million (kilowatt-hours actually used) divided by 876 million, or 57 percent.

The annual load factor is found by dividing the kilowatt-hours delivered during the year by the kilowatt-hours that would have been delivered if the maximum demand during any one hour had prevailed every hour, every day of the year.

3

Edison Power Company

The Edison Power Company is a hypothetical company that
will be used to demonstrate the principles that apply to the
electric power industry. The company was formed when two
smaller companies merged. These companies were the Rosedale Edison Company, which served the towns of Rosedale,
Sharpsville, and Crystal Lake, and the Pleasantville Electric
Light and Power Company, which supplied electricity to the
towns of Pleasantville and nearby Carbon City. The companies were merged to take advantage of greater economies
through unified operation. These economies are discussed in
later chapters. Edison Power Company's service area includes
all five of these towns. Chart 3.1 shows the location of some
of the company's generating plants and the transmission lines
connecting them.

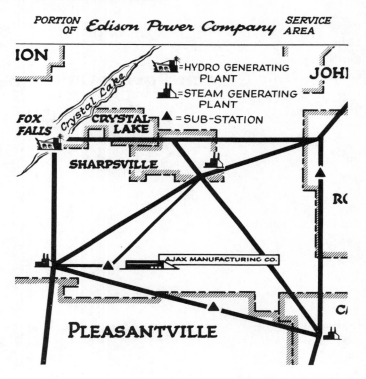

PORTION OF *Edison Power Company* SERVICE AREA

ION

Crystal Lake

= HYDRO GENERATING PLANT

= STEAM GENERATING PLANT

= SUB-STATION

JOHI

FOX FALLS

CRYSTAL LAKE

SHARPSVILLE

RC

AJAX MANUFACTURING CO.

C/

PLEASANTVILLE

CHART 3.1

Physical Plant

Edison Power Company generates most of its electricity by means of coal-burning steam plants, located for the most part near the major *load centers*—places where considerable electric power is used. In addition, the company has one hydroelectric (or water-power) generating station. This station takes advantage of favorable water conditions at Fox Falls to make electricity from the power of falling water.

Generation System. The energy in the fuel or in the falling water is converted to electrical energy at the *generating sta-*

tions or power plants. A power plant may have one or more turbogenerators. The turbogenerator, sometimes called a *unit* in the plant, is a steam- or water-driven turbine directly connected to an electric generator. The turbine is turned by steam in the steam plants and by falling water in a "hydro" plant. Generators are rated in terms of kilowatts of guaranteed continuous output.

The map shows four of Edison Power Company's generating stations. One is located at the only available water-power site. The steam plants are placed at strategic points with respect to the power requirements of the territory served. In locating these plants, the company has considered the nearness of the fuel supply and the availability of water which is used in a steam plant for steam condensing purposes.

The power plants must have *reserve* or *standby capacity* so that the company's customers will still have power even if one of the units fails. The reserve may be either a spare generating unit, or the company may have an interconnection with a neighboring company so that it can obtain power in an emergency. Edison Power has both its own reserve and an interconnection with a neighbor company. Edison's reserve capacity consists partly of older generators. These are kept in tiptop condition ready to go, but stand idle most of the time because it costs more to run these older plants than it does to operate the newer ones. The capacity of the older plants is ready for use in an emergency or when people want to use more electricity than the other plants can produce. This does not happen very often.

If the reserve unit is in operation (that is, spinning at full speed), it is called a *spinning reserve*. A machine that is running can generate power almost instantaneously. The spinning reserve is running, but often is not generating any electricity, and is using only enough fuel to overcome losses caused by friction. At times a boiler is kept hot, ready to operate a turbine, but the turbine may not be spinning. This is called a *hot reserve*. A reserve generator with no heat under the boiler is called a *cold reserve*. It may take an hour or more to bring the machine from a cold reserve to a spinning reserve, or to a condition where it can generate power.

The sum of the capacities of all units in the power plant is termed the plant capacity and is usually expressed in kilowatts. (Capacities, especially of larger units and plants, are frequently expressed in *megawatts*. A megawatt is 1,000 kilowatts.) The sum of all the plant capacities and the capacities which are purchased from other companies is called the *system capacity*. This is also expressed in kilowatts. The total capacity of the system minus the capacity of the largest single unit in the system is called the *firm power capacity* of the system. This *firm capacity* is the highest load the system can carry in the event that the largest unit should break down. (Some very large systems may have more than one unit as reserve.) To provide for this contingency, the reserve capacity must be at least equal to the size of the company's largest unit.

A system with several power plants will make the greatest use of its generating units with the lowest production cost. The company will likely run its lowest-cost units all the time, day and night, to carry all the load they can carry. A unit so loaded is spoken of as a *base load unit*. If the demand increases above the capacity of the base load units, the next most efficient unit is put in operation.

Transmission System. Chart 3.2 shows a diagram of Edison Power's steam plant near Pleasantville. Right next to the power plant are step-up *transformers*. These transformers take the electricity from the generators and raise it to a higher voltage for transmission over long distances. This group of transformers is referred to as a *step-up substation*. The step-up transformers serve the same purpose as a water pump at the beginning of a water pipeline; the pump boosts the water pressure so that it can send the water over a long distance rapidly.

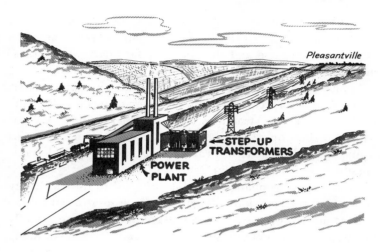

Pleasantville

←STEP-UP
TRANSFORMERS

POWER
PLANT

CHART 3.2

A line used solely for transmitting power from a generating plant to a distant center of load, or from one load center to another, is a transmission line. The group of interconnected transmission lines that carries the energy at a high voltage is termed the transmission system. Transmission voltages usually range from 69,000 to 500,000 volts. Currently field tests are planned using voltages of 1 million volts and above. These voltages will be used to handle the increased amounts of power that the growth of the industry will soon require. They will also permit the economical transportation of power over greater distances. Edison Power as a moderate-sized company has found that 230,000 volts is adequate for its major transmission lines.

Transmission lines emanating from power plants usually form a network looping through various communities and other power plants. Chart 3.1 shows this arrangement of Edison Power's transmission system. No matter where a break in the line occurs, electricity can still be sent to any of the communities by an alternate route while the broken line is being repaired.

Chart 3.3 shows the main high-voltage transmission lines in the United States. In the Rocky Mountain region where the population is sparse and the cities are far apart, fewer transmission lines are needed.

Transmission lines are either aluminum or copper wires, called *conductors,* carried on steel towers or wood pole structures. Wood pole structures may be either a *single pole* or what is called an *H-frame* structure composed of two poles connected by a crossarm near the top. Some typical transmission towers are shown in Chart 3.4.

As motors or appliances are not designed to use electricity at the high transmission voltages, it is necessary to reduce the voltage at load centers. These voltage reductions are usually made in each community served or at some other point along the transmission line to serve a very large power customer. The transformer substations used to lower the voltage from the transmission voltage are called step-down substations. (As a rule, the term "substation" when used alone signifies a step-down substation. The step-up substation usually has the prefix.)

CHART 3.3

BACKBONE TRANSMISSION MAP·1969

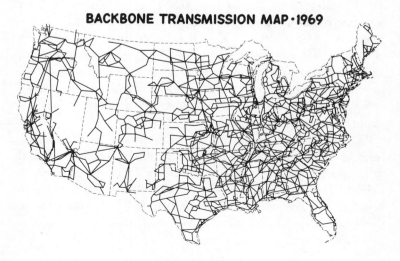

TRANSMISSION TOWERS

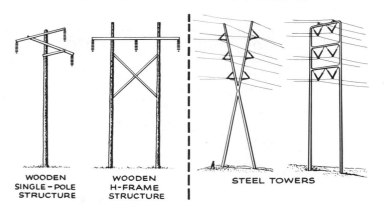

WOODEN
SINGLE - POLE
STRUCTURE

WOODEN
H-FRAME
STRUCTURE

STEEL TOWERS

CHART 3.4

Chart 3.1 shows that Edison Power has a substation just outside of Pleasantville, to provide power at usable voltage for the Ajax Manufacturing Company.

The *main transmission system* of bulk power supply is the system of transmission lines interconnecting a company's major power plants and the lines interconnecting with other power suppliers. Thus, the transmission system for bulk power supply is distinguished from the lower voltage transmission lines serving certain communities and industries and from the low-voltage distribution systems within the communities.

Distribution System. Chart 3.5 shows how a step-down substation near Rosedale lowers the voltage from 230,000 to 13,-200 volts. The lines carrying electricity at 13,200 volts are called *primary distribution lines,* and they extend throughout the area in which electricity is to be distributed. This voltage is still too high for use in the home or factory, so it is reduced again by a *line transformer* or *distribution transformer.* The distribution transformer changes the voltage from the primary distribution voltage (this voltage may vary from company to company; Edison Power uses 13,200 volts) to the

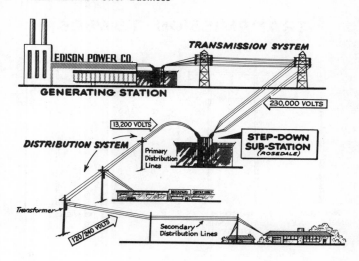

CHART 3.5

secondary distribution voltage, which for use in the home is usually 120 or 240 volts. For the use of motors in a factory, the voltage may be 240, 480, or 2,400 volts.

The lines that run to the customer's premises are called *services* or *service wires*.

The combination of all the primary distribution lines, distribution transformers, secondary distribution lines, and services is called the *distribution system*.

How the Industry Grew

The first electric generating plants and distribution systems were located in the principal cities. In those days, there was no economical way to transmit electric power over any great distance. Later, as ways to transmit electricity over many miles were perfected, the electric companies were able to bring electric service to the smaller communities. The next step was to connect the power plants of one city with the power plants of another city, which permitted many more economies. As transmission lines were extended, larger electric generating plants could be built, and they could be located nearer to supplies of fuel and water. In the course of time, the electric companies built a rather complete network of transmission lines covering most of the country. Electric power companies not only interconnected their plants within their own systems, but they also interconnected their systems with those of other companies.

After the transmission lines had been extended to the small towns and hamlets, service was further extended into rural and farm areas. Because of the distance between farms as compared with the distance between residences in the city, farm electrification developed more slowly than urban electrification.

In the late thirties, great strides were made in this final phase of building the nation's network of power lines. Today, practically every village, hamlet, city, and town in the country has electric service, and the rural electrification program is virtually complete so far as line extension to farm customers is concerned. (For a mention of rural electric cooperatives see page 83.)

How the Company Is Financed

Edison Power Company has $777 million invested in plants, facilities, and other assets as shown in Table 3.1 (see also Chart 3.6).

Money for this kind of investment is raised (1) by selling securities to the public—individuals, banks, insurance com-

TABLE 3.1   Investment, Edison Power Company

Generating plants.	$273,000,000
Transmission facilities.	125,350,000
Distribution facilities.	269,000,000
Miscellaneous property.	94,850,000
Total plant and equipment.	$762,200,000
Other assets.	14,985,200
Total investment.	$777,185,200

Edison Power Company INVESTED CAPITAL

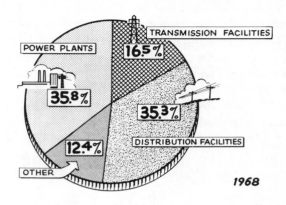

CHART 3.6

panies, pension trusts, and the like; (2) from earnings which a company is able to retain from its operations; and (3) from other internal cash sources, such as depreciation charges.

A company usually does not pay out all of its available earnings in dividends to its stockholders. It retains a portion of its earnings to provide against the possibility that it may not earn as much in some years as in others. Although this money belongs to the stockholder, he does not receive it, but allows the company to invest it in new facilities. The book value of the common stock is thereby increased for the stockholder, and the stockholders as a rule eventually receive an increase in their dividends as the company grows.

Depreciation is a charge made against a company's operations over the lifetime of a facility to provide for the anticipated replacement of the facility when it wears out. Since there is no immediate need for the money represented by these charges, the cash is put to use by the company for building other facilities until it is needed for making replacements. Depreciation is further discussed on page 55.

Edison Power Company obtained its funds from the sources shown in Table 3.2.

Balance Sheet. The *balance sheet* of a company shows

TABLE 3.2   Source of Funds, Edison Power Company

Source	Receipts	Percent
From sale of first mortgage bonds	$300,234,100	38.6
From sale of preferred stock	52,507,100	6.8
From sale of common stock	169,325,255	21.8
From retained earnings.	43,526,145	5.6
From depreciation charges, etc.	211,592,600	27.2
Total	$777,185,200	100.0

what a company owns, what it owes, and what the stockholders' interest in the company is. A balance sheet is commonly set up in two columns. The left-hand column shows what the company owns. This is called *assets,* and consists, in the main, of *fixed assets*—plant and equipment—and *current assets*—cash, accounts receivable, materials and supplies, etc., which usually can be turned into cash readily or are used up in the operation of the business. The right-hand column of a balance sheet shows what a company owes, called *liabilities,* and the stockholders' interest in the company, which is often referred to as the *stockholders' equity.* Liabilities consist of the long-term liabilities, such as mortgage bonds and other long-term debt, *current liabilities,* and *reserves.* Obligations that do not become due for over one year are usually considered long-term debt. Current liabilities include accounts payable, notes that are due within a year, payroll, accrued taxes, and the like. The difference between current assets and current liabilities is sometimes referred to as *net current assets.* The relationship between current assets and current liabilities is called the *current ratio.*

Reserve accounts record amounts charged against earnings for future payments or contingencies, such as depreciation reserves. These accounts are, for the most part, "bookkeeping" accounts and do not represent cash reserves necessarily.

The stockholders' equity includes preferred and common stock which the company has outstanding and accounts called *Capital Surplus* and *Earned Surplus.*

Preferred stock usually has a par value and is stated on the balance sheet at this value. This is the amount which the company is obligated to pay the preferred stockholders in the case of a voluntary dissolution.

Common stock may have a *par value* or it may be *no-par stock*. In the case of stock having a par value, this value is reported for the common stock outstanding, and any amounts received by the company from the sale of stock in excess of par value are reported as capital surplus. Capital surplus also includes amounts such as may have been received from the sale of assets in excess of book value and other amounts which are usually the result of corporate transactions.

Earned surplus represents the accumulated earnings of the company, after adjustments for such things as over- or under-accruals, less dividends paid to the stockholders.

The stated value of the common stock plus the capital surplus and the earned surplus represent the common stock equity. The sum of these is the book value of the common stock. Sometimes amounts in the Capital and Earned Surplus accounts are transferred. Amounts in Earned Surplus can be transferred to Capital Surplus, or amounts in either Earned Surplus or Capital Surplus can be transferred to the stated value of common stock, as in the case when stock dividends are issued by the company. Such transfers do not change the book value of the stock, however.

Since the assets of a company always equal the liabilities plus the stockholders' equity, the statement of these items is called a balance sheet.

A condensed balance sheet of Edison Power Company at December 31, 1968, is shown in Table 3.3.

Stocks and Bonds. The mortgage bondholder has first claim on the assets of the company and thus has the highest degree of security. Next in line, in order of security, after other creditors, is the *preferred stockholder*. While a bond represents, in effect, a loan to the company which it must eventually repay, *preferred* and *common stock* represent shares of ownership of the company. Edison Power's preferred stock

TABLE 3.3   Condensed Balance Sheet, Edison Power Company

Assets		Liabilities	
Electric utility		Common stock . . .	$139,325,255
plant	$762,200,000	Preferred stock . . .	52,507,100
less:		Capital surplus . . .	30,000,000
Reserve for		Earned surplus . . .	43,526,145
depreciation . .	178,592,600		
Net electric		Total capital stock	
utility plant	$583,607,400	and surplus	$265,358,500
Investments and		Long-term debt . . .	300,234,100
other assets	14,985,200		
		Total capitalization .	565,592,600
Current assets	42,000,000	Current liabilities . .	50,000,000
		Other reserves	25,000,000
		Total	
Total assets . .	$640,592,600	liabilities . .	$640,592,600

has first claim on the earnings of the company after bond and other interest is paid, so that its dividend rate is comparatively secure. It also has first claim—after the bonds, and other creditors, of course—on the assets of the company, so that the preferred shareholder thus assumes less of a risk than the common shareholder.

Table 3.4 shows how the preferred stock of Edison Power Company is distributed.

TABLE 3.4   Distribution of Preferred Stock, Edison Power Company

Owners	Number of shares	Percent
Women	1,422,000	35.6
Men	896,000	22.4
Joint owners	457,000	11.4
Fraternal, charitable, religious, and educational funds	54,000	1.3
Trusts and custodian accounts	355,000	8.9
Insurance companies.	810,000	20.2
Dealers, corporations, and others	6,000	0.2
Total	4,000,000	100.0

The *common stockholders* are the real owners of the company. They assume the greatest risk of loss, and if the company fails to earn enough money to pay bond interest, preferred dividends, and a return for common stock, the common stockholders suffer. On the other hand, if the company does well, the common stockholders stand to benefit, and have a greater chance to benefit than holders of bonds or preferred stock. In actual practice, government regulation prevents spectacular common stock benefits in the electric power business. But regulation does not guarantee that the company will necessarily make a profit.

Table 3.5 shows how the common stock of Edison Power Company is held. The percentage distribution is illustrated in Chart 3.7.

TABLE 3.5   Distribution of Common Stock, Edison Power Company

Owners	Number of shares	Percent
Women	4,500,898	34.5
Men	3,378,935	25.9
Joint owners	2,335,248	17.9
Fraternal, charitable, religious, and educational funds	313,106	2.4
Trusts and custodian accounts	1,408,977	10.8
Insurance	221,783	1.7
Dealers, corporations, and others	887,133	6.8
Total	13,046,080	100.0

The *capitalization* of the Edison Power Company consists of its long-term debt, preferred and common stock outstanding in the hands of the public, and capital and earned surplus, as shown in Table 3.6.

The manner in which a company is capitalized has an important bearing on how its securities are rated in the money market and, therefore, the cost to the company of raising additional funds.

Securities offered on the public market are rated by professional rating agencies such as Moody's Investors Service or

DISTRIBUTION of COMMON STOCK
Edison Power Company

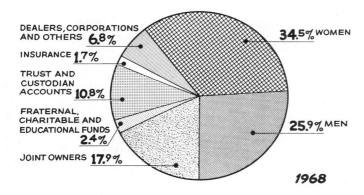

DEALERS, CORPORATIONS AND OTHERS **6.8%**

INSURANCE **1.7%**

TRUST AND CUSTODIAN ACCOUNTS **10.8%**

FRATERNAL, CHARITABLE AND EDUCATIONAL FUNDS **2.4%**

JOINT OWNERS **17.9%**

34.5% WOMEN

25.9% MEN

1968

CHART 3.7

Standard and Poors. The purpose of the rating is to provide the investor with a simple guide by which the relative investment qualities of bonds may be noted. These ratings are assigned to a company's securities after exhaustive studies of the company's operations, financial ratios, management, territory served, market characteristics, past performance, regulatory climate, etc. Ratings range from Aaa (Moody's), AAA (Poors), which are judged to be of the best quality carrying the smallest degree of investment risk, to C, which are regarded as having extremely poor prospects of ever attaining any real investment standing. Most electric utility company

TABLE 3.6   Capitalization, Edison Power Company

Source	Amount	Percent
First mortgage bonds	$300,234,100	53.1
Preferred stock	52,507,100	9.3
Common stock	139,325,255	24.6
Capital surplus	30,000,000	5.3
Earned surplus (retained earnings) . . .	43,526,145	7.7
Total	$565,592,600	100.0

first mortgage bonds are rated A or better by these services. A number are triple A and many are double A. An A-rated bond is considered to be a high medium-grade obligation giving adequate security to principal and interest; a double A rating indicates a higher quality but not quite as "gilt edge" as triple A.

A rating generally is reflected in the relative interest cost of money; that is, in a given market, a triple-A-rated electric utility bond will generally sell at a lower yield than a double A, and a double A at a lower yield than an A-rated bond. Particular market conditions and terms of the contract influence the spread in yields among bonds of the different ratings.

Edison Power Company tries to see that bonds make up not much more than 60 percent of its total capital structure, and its common stock equity is between 30 and 40 percent. This helps the company keep a good credit rating and maintains its ability to raise new funds when needed. The ratio of its equity to the total capitalization also has a bearing on how people evaluate the company's common stock.

The principal *bondholders* of Edison Power Company are institutional investors. They represent a group of investors or savers whose savings they invest in securities of various types of business. These institutional investors include insurance companies, investment trusts, pension funds, banks, and the like. For example, on January 1, 1968, life insurance companies held nearly half of the long-term bonds of the company. Insurance companies invest part of the premiums paid by their policyholders in securities. The money earned on these securities is used to reduce the amount that policyholders have to pay in premiums for their insurance. In this respect, the people who have insurance policies with the investing companies are indirect investors in the Edison Power Company.

Principal Sources of Income

Edison Power Company, like most other electric utility companies, receives most of its revenue from sales at retail to residential, commercial, and industrial customers.   Chart 3.8 shows the number of customers in each of these classifications and the percentage of the total each class represents.   The company has 544,187 customers.

Chart 3.9 shows the company's annual revenue by classes of service in percent of total revenue.   The company had a total revenue of about $161,391,194 in 1968.

Expenses of Operation

So far, mention has been made of only one item of cost—the interest or return that a company must pay investors for the use of their money.   There are, of course, other expenses. Those following are referred to as operating expenses.   Chart 3.10 shows where the company's money goes.

CHART 3.8

Edison Power Company **CUSTOMERS**

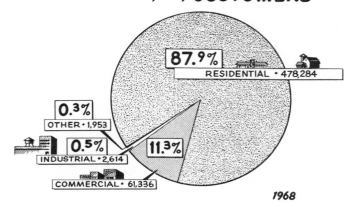

87.9% RESIDENTIAL • 478,284

0.3% OTHER • 1,953

0.5% INDUSTRIAL • 2,614

11.3%

COMMERCIAL • 61,336

1968

Edison Power Company ANNUAL REVENUE
BY CLASSES OF SERVICE
INDUSTRIAL
24.1%
OTHER
10.6%
COMMERCIAL
27.8%
37.5%
RESIDENTIAL
1968

CHART 3.9

Production Expense. The principal items of expense in running a power plant are fuel, labor, material, and supplies. The total of all of these is called the production expense. For Edison Power, the production expense in 1968 was \$40,900,000, or 0.378 cent (3.78 mills) per kilowatt-hour sold.

CHART 3.10

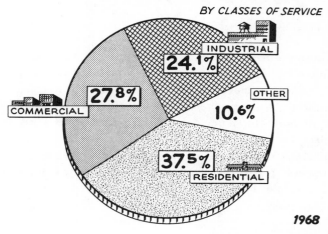

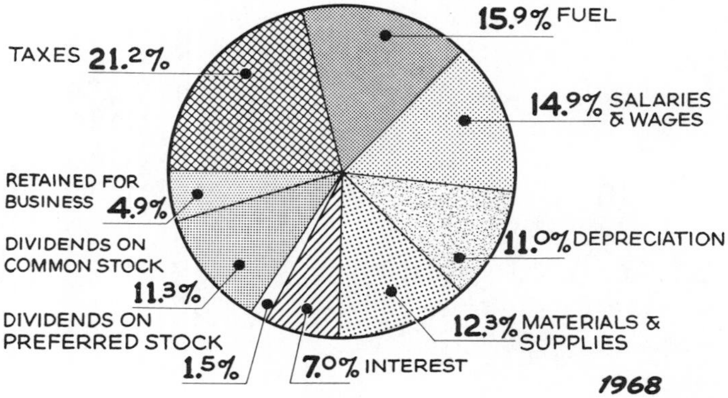

WHERE THE MONEY GOES...
*As a percent of gross operating revenue
Edison Power Company
15.9% FUEL
TAXES 21.2%
14.9% SALARIES & WAGES
RETAINED FOR BUSINESS 4.9%
11.0% DEPRECIATION
DIVIDENDS ON COMMON STOCK
11.3%
DIVIDENDS ON PREFERRED STOCK
1.5%
7.0% INTEREST
12.3% MATERIALS & SUPPLIES
1968

(Note that some expense items are given in cents per kilowatt-hour, while others are given on other bases, such as dollars per customer. These figures are given in this way here to give an idea of the size of the item. This method of expressing the data has a useful purpose, which is described in Chapter 9, "Costs and Pricing.")

Edison Power Company uses coal as fuel in its steam plants. On the average, Edison Power's plants require about 0.87 pound of coal to make each kilowatt-hour. The company used 4,281,778 tons of coal in 1968 which cost it $25,-690,665, or 15.9 percent of the company's total revenue. This cost amounted to 0.261 cent (2.61 mills) per kilowatt-hour generated.

While the fuel expense is an important item, it is not nearly as big as some other costs of furnishing electric service. For example, it is not as large as the return the company must pay investors for the use of their money, and it is not as large as taxes.

With the coming of atomic energy it is contemplated that the fuel element in the over-all cost of making electricity will continue its gradual downward trend. In all probability there will not be a spectacular drop over the years, but the effect will be reflected in the price of electricity.

Frequently, a company has a clause in its rate schedules providing for a corresponding increase or decrease in the rate for any increase or decrease in the cost of fuel. (Costs and pricing are discussed in Chapter 9.)

Transmission Expense. Transmission expense includes wages of people who work on the high-voltage lines, keeping them in good order, and the expense of inspecting the transmission grid at regular intervals.

Distribution Expense. This expense item covers the cost of labor to operate and maintain the distribution lines, substations, and other facilities. Money paid for materials used in maintenance and the expense of patrolling, inspection, and testing is also included.

Customer Accounts Expense. This category contains the cost of reading meters, making out and mailing customers' bills, keeping the accounts, and collecting bills. For Edison Power Company this expense amounted to $7.52 per customer in 1968.

Sales Expense. Sales expenses include such things as advertising, wages, and expenses to obtain new business, cost of showrooms, and allied items.

Administrative and General Expense. This account covers the salaries and expenses of general officers, executives, and general office employees. It also includes insurance, employee benefits, and items of expense not specifically provided for elsewhere. This expense amounts to 5.9 cents per dollar of gross revenue.

Taxes and Depreciation

Taxes. In addition to the operating expenses, the company must also take into account the effect of taxes and depreciation before it can arrive at the *balance for return,* the amount available to pay the investors for the use of their money.

The electric utility company is required to pay many kinds of taxes. Among these are local property taxes, franchise taxes, excise taxes, and state and Federal income taxes.

Edison Power Company had a total of $34,200,000 in taxes last year. Of these total taxes $18,416,349 were for Federal income taxes, which amounted to 11.4 cents per dollar of gross revenue, or 2.4 percent of the total value of the company's plant investments.

The total tax bill took 21.2 cents out of every dollar the company received from its customers. This is more than the company paid for fuel and more than the company paid for labor. It was also more than the company paid to its stockholders. It amounts to 4.5 percent of the total cost of the company's plant investments.

Chart 3.11 shows the taxes of Edison Power Company

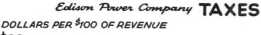

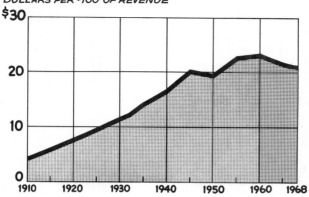

Edison Power Company **TAXES**

DOLLARS PER $100 OF REVENUE

CHART 3.11

through the years. It shows the taxes per $100 of annual gross revenue for typical years since 1910.

A company has no income except the money it receives from customers for the commodity or product sold by the company. Taxes, like labor and fuel, are an expense and, like other costs, are included in the price of the commodity. Thus, in effect it is the customers who pay the company's taxes.

Depreciation. A utility company must keep its property in good working condition to maintain continuous satisfactory service to the public. This means that equipment must be replaced when it wears out, becomes obsolete or inefficient, or is taken out of service for any reason. When property is taken out of service, it is retired from the company's property account. When this retirement takes place, the company must be in a position to replace the retired property with new equipment.

It is not possible to predict exactly when it will be wise or necessary to retire any particular piece of property. Each year, the company reserves from its income an amount to account for the estimated cost of property wearing out. The accumulation of these charges less the amount deducted for property retired is called the *reserve for depreciation,* or *depreciation reserve.*

Edison Power Company set aside $17,700,000 in 1968 as an accrual to the depreciation reserve for its depreciable property. This annual charge is equal to 2.3 percent of the company's total investment in physical property.

Net Operating Revenue

The *net operating revenue* (or *operating income,* as it is called in the revised classification of accounts) is the amount left over from gross operating revenue after payment of all operating expenses, depreciation, and taxes. It is the amount of money left from operations to compensate the investors for the use of the funds they have entrusted to the company.

Most companies have some other incidental income which when added to net operating revenue results in *gross income.*

For sake of simplicity it is assumed Edison Power Company has no such incidental income. Consequently net operating revenue is the same as gross income.

The company sends an annual report to its security holders and others who may be interested in the company's operations. This report details the amount of money the company received during the year from its customers, and gives the amounts paid out in expenses. An extract from Edison Power Company's Annual Report for 1968 giving certain revenues and expenses is shown in Table 3.7.

The net operating revenue of Edison Power Company was distributed in 1968 as shown in Table 3.8.

The money that is left over after payment of bond interest and preferred dividends is the share accruing to the common stockholders. However, good management does not pay all of this in dividends; the company usually pays around 65 to 75 percent of it in dividends to the common stockholders. The balance is retained as a reserve for unforeseen contingen-

TABLE 3.7   Revenue and Expenses, Edison Power Company

Item	Amount	Percent of gross operating revenue
Gross operating revenue.	$161,391,194	100.0
Operating expenses:		
Production	40,900,000	25.3
Transmission	2,110,000	1.3
Distribution	10,100,000	6.3
Customer Accounts	4,090,000	2.5
Sales	2,905,000	1.8
Administrative and general	9,500,000	5.9
Total operating expenses °. . .	$ 69,605,000	43.1
Depreciation	17,700,000	11.0
Taxes	34,200,000	21.2
Total operating expenses, . . .		
depreciation, and taxes . . .	$121,505,000	75.3
Net operating revenue	$ 39,886,194	24.7 †

° The relationship between operating expenses and annual gross revenue is sometimes referred to as the operating ratio. For Edison Power Company, the operating ratio is $69,605,000 divided by $161,391,194, or 43.1 percent.

† This should not be confused with percent return on investment. The $39,886,194 represents 5.2 percent of the book value of the plant and approximately 7.1 percent of total capitalization.

cies and for the expansion and improvement of the business. This amount is reflected in earned surplus unless capitalized by issuance of stock dividends, transfers to capital surplus, or restatement of common stock stated value. It has the effect of increasing the stockholders' equity in the business.

TABLE 3.8   Distribution of Net Operating Revenue, Edison Power Company

Distribution	Amount	Percent
Bond interest	$11,299,860	28.3
Preferred stock dividends	2,494,170	6.3
Common stock dividends	18,264,515	45.8
Retained in the business	7,827,649	19.6
Total	$39,886,194	100.0

WHERE THE MONEY GOES*...

Edison Power Company

DEPRECIATION and AMORTIZATION 11.0%

21.2% TAXES

ADMINISTRATIVE and GENERAL 5.9%

SALES 1.8%

CUSTOMER ACCOUNTS 2.5%

DISTRIBUTION 6.3%

TRANSMISSION 1.3%

PRODUCTION 25.3%

OPERATING EXPENSES 43.1%

NET OPERATING REVENUE 24.7%

1968

**AS A PERCENT OF GROSS OPERATING REVENUE*

CHART 3.12

Chart 3.12 and Chart 3.13 show how expenses are apportioned and the distribution of return to the investors in Edison Power Company.

CHART 3.13

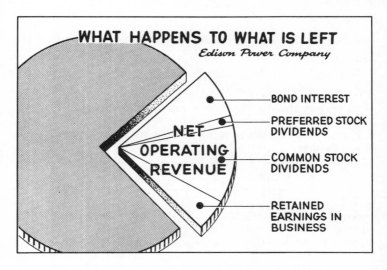

WHAT HAPPENS TO WHAT IS LEFT

Edison Power Company

NET OPERATING REVENUE

BOND INTEREST

PREFERRED STOCK DIVIDENDS

COMMON STOCK DIVIDENDS

RETAINED EARNINGS IN BUSINESS

Other Operating Data

Kilowatt-hour Sales. Chart 3.14 shows the company's annual sales in kilowatt-hours by classes of service and the relative size of each class. Total sales last year amounted to about 11 billion kilowatt-hours.

Loads. Understandng the loading on the power system helps in gaining an understanding of the economics of the power business.

Chart 3.15 shows a typical daily load curve for the residential class of service of Edison Power Company for a day in the wintertime. For each of the twenty-four hours, it shows the number of kilowatts being demanded by the company's residential customers.

Starting at midnight the demand is light. Refrigerators are off and on all night, and certain types of water heaters are heating water. Beginning about 5 A.M., however, the load increases. Lights begin to go on, and electric ranges are being

CHART 3.14

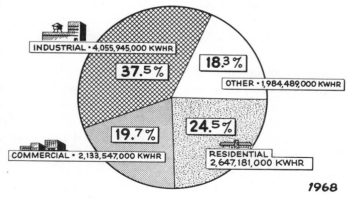

ANNUAL SALES in KWHR
Edison Power Company

INDUSTRIAL · 4,055,945,000 KWHR

37.5%

18.3%

OTHER · 1,984,489,000 KWHR

19.7%

24.5%

COMMERCIAL · 2,133,547,000 KWHR

RESIDENTIAL 2,647,181,000 KWHR

1968

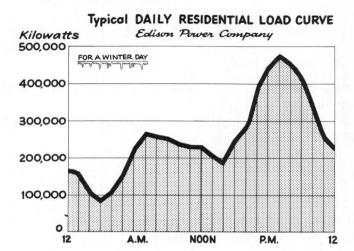

CHART 3.15

used to cook breakfast. Later there is laundry to do, dishwashing that takes hot water, and perhaps some electricity is used in watching soap operas on television. At noon the demand is about 235,000 kilowatts. Some companies that have relatively large industrial loads have a peak in the morning higher than the one in the evening. For Edison Power Company, the load rises to its highest peak of the day around 7 P.M. Most lights are on then, and housewives in the area are preparing the evening meal. On a summer day, the peak may occur at about 3 o'clock in the afternoon, when air conditioning demand is at its greatest.

The company has a load dispatcher who is constantly watching the company's control dials that show the variation in load on the company's system. Because he has spent years on this job, he can predict fairly accurately what the load will be each hour. He keeps in constant touch with the plant operators, telling them which generators to start in operation. The trend is toward greater use of electronic computers in this operation.

Comparison of the daily load curves from day to day shows

that, although the general pattern is the same, the actual load varies from day to day and from month to month. However, it is the peak, or highest load, which is most important, since this determines the amount of generating capacity the company must provide. It takes a long time to plan and build generating units. For this reason, the company must plan ahead and predict the peak load for a number of years in the future.

Chart 3.16 shows the highest peak load for each of the twelve months. This is called the annual load curve. The chart shows that Edison Power Company has its highest peak load in December. Several years ago, the December peak was much more pronounced, but the popularity of air conditioning has caused a rise in the load during the summer months. Now the December peak exceeds the summer peak by only a small margin.

In order to demonstrate the peaks and valleys in the company's load curve, Chart 3.17 shows a three-dimensional view

CHART 3.16

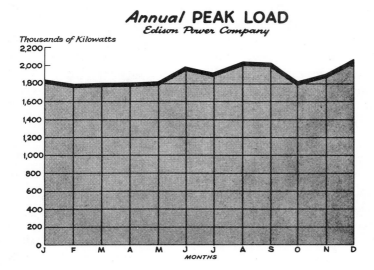

ANNUAL LOAD
Edison Power Company

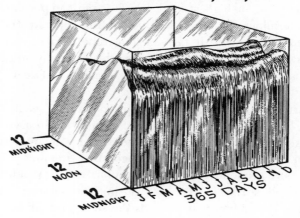

CHART 3.17

of the way Edison's customers use electric service during the course of the year.

The highest peak is 2,056,000 kilowatts. If the company's customers had used electricity at this rate during every hour of the year, the company would have generated 18,011 million kilowatt-hours (2,056,000 kilowatts × 8,760, the number of hours in a year). Allowing for system losses, the company could have sold 16,570 million kilowatt-hours. The company actually sold 10,821 million kilowatt-hours during that time. Its load factor was 65 percent (10,821 ÷ 16,570).

Use of Electricity Increasing. The uses for electricity are constantly increasing. While there is less of an increase in kilowatt-hour use during times of depression, on the whole the trend is upward and will probably be so for many years to come. The increase is at the rate of about 7 percent a year. The load served by Edison Power Company is increasing at this rate, as shown in Chart 3.18.

The chart shows Edison Power's peak load by years since 1915. Note the leveling during the depression of the early thirties and also the increase since World War II. New generating capacity has been built to keep ahead of the load.

Units in the Plant. The largest unit in the company's Pleasantville power plant is a 500,000-kilowatt generator. The company can use units of this size because it is a part of a power pool (see Chapter 11). Larger power companies can probably make economic use of units much larger than this.

In engineering practice, several terms are used to designate the rating of a generating unit. *Capacity* or *name-plate rating* refers to the manufacturer's rating of the unit. The engineers who design the unit draw up their specifications so that the finished generator will produce electricity at a certain rate under specified conditions. The manufacturer might say of the finished product that it will generate electricity at the rate of 500,000 kilowatts, for example. He will imprint this rating on the name plate attached to the generator, along with the design conditions. This rating is given in gross kilowatts—that is, without any deduction for electricity used in the generating process (such as for operation of fuel pumps and other auxiliary purposes).

CHART 3.18

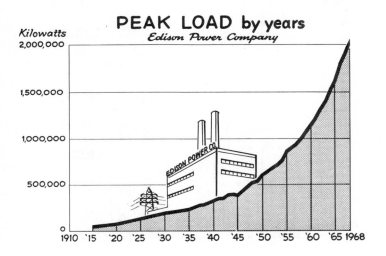

PEAK LOAD by years
Edison Power Company

Capability is another term used in rating generators. After the company has had the 500,000-kilowatt unit in operation for a period of time, it may find that because of favorable operating conditions the unit is actually capable of delivering electricity at the rate of, say, 515,000 kilowatts. This is called "capability." Capability is given in net kilowatts, after deduction for electricity used in the generating process. After the unit is installed it is more realistic to deal in terms of capability instead of capacity.

In nontechnical language, the word "capacity" is a generic term which includes capability. It is frequently so used in this book.

The company's capability has gone up in stair-step fashion as shown in Chart 3.19. When the company builds a plant, it leaves ample space for additional units to be installed in future years. This tends to lower the cost per kilowatt by making more efficient use of the site, buildings, and transmission lines, and also means that the company does not need to

CHART 3.19

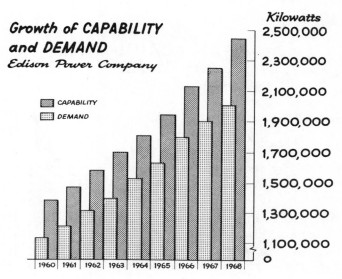

Growth of CAPABILITY and DEMAND

Edison Power Company

CAPABILITY

DEMAND

Kilowatts

2,500,000

2,300,000

2,100,000

1,900,000

1,700,000

1,500,000

1,300,000

1,100,000

0

| 1960 | 1961 | 1962 | 1963 | 1964 | 1965 | 1966 | 1967 | 1968 |

build a new plant every time the demand rises. When the demand requires a new unit to be built, the company usually puts in the largest practical unit. It is likely that after a new unit has been added, the company will have, for a while, more capacity than it actually needs. This excess capacity may be sold to neighboring companies until it is needed by Edison Power Company.

With the development of more coordination and pooling there frequently is joint planning with more than one company on new generating capacity. Thus maximum advantage can be taken of the economies to be derived from large generating units.

Other Capacity. Just as the generating capacity must be increased as the customers' demands for service increase, so also must the capacity of other equipment be increased. A step-up transformer is added at the power plant whenever a new generating unit is installed. The transmission lines, the power substations, the distribution systems—all must be increased in capacity as the rate of use of electricity increases. Always, all along the line, the customers' requirements must be anticipated and provided for in advance.

Personnel Organization and Payroll

Edison Power Company employs 3,900 people. As the company has invested $762,200,000 in plant and equipment, this is an investment of $195,000 per employee. Because of the extensive use of machinery and the technical nature of the business, utility employees are usually more highly skilled —and better paid—than ordinary employees.

Of the total payroll of $24,000,000, the salaries of management amount to $768,000, or 3.2 percent of the total payroll.

The stockholders, who are owners of the company, elect the Board of Directors. The Board selects the President, who is the company's chief executive officer, and as such the Board holds him responsible for all of the company's operations.   In some companies the Chairman of the Board of Directors acts as the chief executive officer.

The President may have reporting to him one or two executive vice presidents and a number of department heads, such as Treasurer, Chief Engineer, Sales Manager, Director of Public and Employee Relations, Rate Engineer, General Counsel, and the like.   Frequently many of these department heads carry the title of vice president.

Power companies vary widely as to the number of divisions they may have.   The territory of the Edison Power Company is divided into four divisions (see Chart 3.20).   For each division there is a Division Manager, who is in charge of all ac-

CHART 3.20

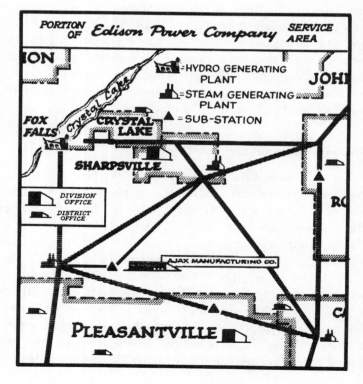

tivities in his division, with the exception of the operation of the power plants and the transmission system. These are operated from the general office. The division itself is organized under the Division Manager into about the same staff functions as the company as a whole. There is, for example, a Division Engineer, a Division Auditor, a Division Sales Manager, and so forth, so that each of the functions is carried on in each division. The company's Sales Manager establishes policy with respect to matters within his scope of operation, and the Division Sales Manager carries out this policy in his division. The same is true for the other staff functions. The Division Manager generally reports directly to the Executive Vice President.

Electric companies usually follow this general pattern, but in detail, the organization varies from company to company.

4

Regulation

Under the American free-enterprise system, competition is generally relied upon to keep prices fair and reasonable. Anyone wanting to buy something can buy it from the seller offering it at the lowest price. People can shop around for their groceries and clothes and automobiles. If a seller sets his prices too high and others are offering the same thing for less money, the higher-priced article simply does not sell. Then the seller must cut his price if he is to continue to make sales. Manufacturers and dealers have to maintain high efficiency in their business operations and must continue to make a good article, reasonably priced, or they cannot stay in business for long.

It is a good system and it works well in most businesses.

But in the case of the electric power business, it was learned early that it did not work to the consumer's advantage. The electric utility business was begun under this competitive system. Companies were formed and operated wherever the investors thought they could sell electricity, and they competed vigorously with one another for customers. An attempt was made to rely upon this competition as a means of keeping the price of electricity fair and reasonable.

For example, in the early days of the electric industry, there were some twenty-five or thirty separate utility enterprises in the city of Chicago. Some were merely isolated generating plants, but others were of fairly good size, using the streets and alleys for distribution lines. They operated at various voltages with many kinds of equipment. There was much duplication of distribution lines. There were no large central stations, and the plants of that day were inefficient as measured by present-day standards.

This condition at times resulted in inferior service, confusion, higher rates to electric customers, and inadequate or no returns to investors. It was also recognized by utility men and government officials that the electric utility business had important differences from usual business enterprises. They realized that, while direct competition was good for business in general, it was costly, impractical, and undesirable both for electricity customers and for electric utility companies.

As a result, the idea of governmental regulation for electric companies, as a fair and workable substitute for competition in kind, began to take form.

This concept recognized, first, that the electric utility has an obligation to serve at reasonable rates all who apply for service; second, that direct competition, in the sense that one grocery or other shop selling tangible commodities competes with another in the same town, was not in the public interest in the case of an electric utility company; and third, that government regulation would provide the safeguard to public interest ordinarily obtained through direct competition.

The states already had experience in regulation. Many states had for years regulated railway companies and gas and water supply. Many cities also had regulatory machinery.

The early lighting companies operated only in cities and towns. Later it became more economical to build large central stations and to send the power over transmission lines. Eventually, a single company would serve large areas and hundreds of cities and towns. This made it desirable to provide statewide regulation. Later, when transmission systems began crossing state lines, Federal regulation of interstate operations came into being.

The Regulatory Bodies

The first states to have formal regulation were New York and Wisconsin. In 1907, public service commissions in these states were given regulatory jurisdiction over electric utilities. Other states followed suit, until today, almost all have means for regulatory control of electric utilities. In forty-six states, public utility commissions watch over the operations of the electric utility companies. (Most also regulate other types of public utility intrastate operations.) In the other states there are local bodies that have regulatory jurisdiction. In Nebraska, almost all electric utilities are owned by the state or local governments and are not regulated by the state.

The Federal government gets its power to regulate electric utilities from the Constitution under the interstate commerce clause and also under provisions for controlling public lands. The power to license water-power sites dates back as far as 1896.

The Federal Power Act was intended to fill a gap in the regulatory process. State commissions had jurisdiction over intrastate matters, but had no jurisdiction over power delivered or received in interstate commerce. The Federal Power Commission, created under the Federal Power Act in 1920, has the power to regulate projects in interstate commerce, on public lands, and on navigable streams and other water resources under Federal control.

The Securities and Exchange Commission has certain authority (Title I of the Public Utility Act of 1935) over holding companies. (A holding company is one which owns the securities of a number of electric companies. It usually operates all of the companies as an integrated system, thereby getting all the benefits of the most modern techniques and equipment.) The SEC also regulates the sale of securities by operating utility companies under the Securities Act. The SEC's control is mainly in the financial field and deals with such things as the limit of total capitalization, capital structure, the terms of security sales, and the protection of utility assets.

The Federal Power Commission has the power to regulate rates for interstate wholesale power, if it finds rates unjust, unreasonable, unduly discriminatory, or preferential. The Colton case, discussed later in this chapter, has broadened the FPC's powers significantly. It may investigate the actual legitimate cost of the property, "the depreciation therein, and, when found necessary for rate making purposes, other facts which bear on the determination of such cost or depreciation, and the fair value of such property" (United States Code Annotated, title 16, sec. 824). In such cases, the quality of service also comes under the commission's jurisdiction. It is empowered to regulate security issues, mergers, and interlocking directorates and managements.

State commissions regulating electric utilities operate under state laws which spell out their duties and powers. They have the authority to look at the records and properties, to question company officials, and to obtain pertinent data from other sources. The commissions hold public hearings, and the operations of these electric utilities are clearly exposed to public view.

State commissions generally have wide powers to decide whether a rate is reasonable.

Why Duplication Is Against
Public Interest

In the normally unregulated business operation there may be as many as five or six intermediate commercial firms between the maker of a product and the person who ultimately uses it (Chart 4.1). The company that makes the product in most cases does not sell it at retail. First, the product is shipped by (1) a separate transportation company such as a railroad or truck line to (2) a distributor or wholesaler, who may store the product and later (3) ship by a local trucking company to (4) the retailer, who may hire (5) a parcel delivery firm to deliver the goods to the consumer.

In the electric power business, one company makes the product and delivers it to the consumer. Electricity is sent over high-voltage expressways to local centers from which it is shipped at a lower voltage to consumers. There is a permanent physical delivery system from the producer of electricity direct to the customer.

CHART 4.1

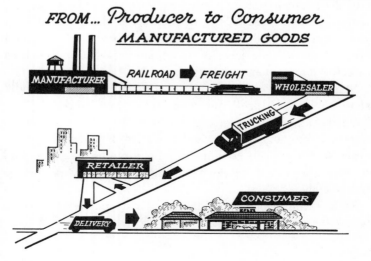

FROM... *Producer to Consumer*
MANUFACTURED GOODS

An ordinary manufacturer can make as much or as little of his product as he desires. He doesn't have to supply all the customers of any particular area. The electric utility company, however, is required by law to meet all, or practically all, the power needs of all the people of a given area.

The manufacturer can store his products. He can run his plant full time and stockpile the product for sale later. Toy manufacturers, for example, run their plants all year round but sell most of their products at Christmas. Such a manufacturer can use his machinery at a fairly steady rate all year. He gets good utilization of the equipment he has had to buy.

But electricity cannot be stored. It is made, delivered, and used at the same instant the customer throws the switch. The company must have generators standing by all the time, waiting for the customer to press the button. All power plants, transmission systems, distribution systems, substations, and transformers must be big enough to meet the diversified demands of all customers who may want to buy electricity at any time.

Some Economic Differences. As the electric utility business is physically unique, there are important economic differences between the electric utilities and other businesses.

For example, because of the high cost of generators and transmission and distribution lines, the electric company has to invest about $4.50 to get $1 of annual gross sales. As a group, manufacturing companies have to invest only about 51 cents to get $1 of sales (Chart 4.2). The investment of the electric utility companies is about nine times that of manufacturing enterprises, to produce the same amount of annual sales.

If two electric utility companies served the same area, the investment and investment costs would be much greater than for a single company. But having two companies in the area would not increase the number of customers, and would not bring about greater sales of electricity. So if there were two companies, the rates to the customer would have to be higher.

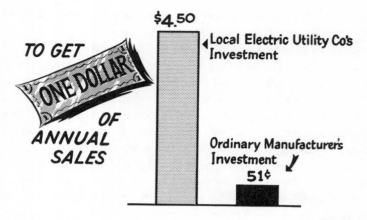

$4.50

TO GET ONE DOLLAR OF ANNUAL SALES

◄ Local Electric Utility Co's Investment

Ordinary Manufacturer's Investment ✔
51¢

CHART 4.2

Meaning of the Term "Monopoly." The word "monopoly" sometimes has a bad connotation in people's minds. They usually think of it as something against the public interest. There are laws against monopolies in certain competitive fields, but these laws do not apply to public utilities which are recognized as natural monopolies and are regulated by government.

There are about 350 independent electric light and power companies in the country, all under some kind of government regulation. In some states, government-sponsored suppliers are not subject to regulation. In other states, they are regulated, but generally not to the same extent as power companies. In areas served by cooperatives, municipal plants, and the like, there is the same kind of monopoly as in areas served by companies. There is seldom duplication.

Competition Continues. Although there is usually only one power company from which a customer can buy electricity, this does not mean that there is no competition for business. There is competition for the customer's dollar. He must be persuaded to buy electric appliances instead of an automobile or a new hat or some other item.

Although there is almost no lighting competition today,

there is competition in other uses, for example, cooking; a customer could use gas or some other fuel. There is competition for water heating, clothes drying, and other domestic uses. As the average price of electricity has decreased, it has become competitive in the large heating market.

There is further competition in the electric power field in that any customer is free to generate his own electricity. Some thirty or forty years ago many industries generated their own power. Their plants were called *isolated generating plants* to distinguish them from interconnected power systems. These isolated plants are sometimes called *on-site generating plants*. In 1920 about 30 percent of the nation's electric energy was generated in these isolated plants. This has gradually come down to about 7 percent as the electric power users have found they can obtain better service at a lower price by purchasing energy from interconnected systems. But the threat of this form of competition continues. Electric power companies must give good service at a low price, as there is always the possibility the customer will make his own power.

In the commercial field there is competition from gas and other forms of heating. In the industrial field there is competition from other forms of energy. A large industrial customer is free to install his own power plant. The company must be able to provide better service or service at a lower cost, or else the customer will not buy electricity.

The people who run a company are constantly trying to demonstrate better management by keeping their rates competitive with others operating under about the same conditions. The companies study new operating methods and techniques in order to improve efficiencies. When a company is able to cut costs and thereby offer lower rates, neighboring companies are eager to learn its methods. Thus there is competition among managements for excellence in performance.

Three Main Objectives of Regulation

Service. Electric utilities usually operate under a franchise or an indeterminate permit which gives them the rights to use streets and alleys and to serve the community and territory. In exchange for these rights the utility undertakes the duty of rendering good and adequate service. The franchises are usually for a fixed number of years. From time to time they must be renewed, and this generally calls for approval by the electorate. Thus the power supplier has a strong incentive to give good reliable service, or else the right to do business in the community might be denied. Of course, the competitive factor in the power business also provides a strong incentive to provide good service at a fair price. The customer can always use other kinds of fuel or generate his own power.

In carrying out its duties in rendering good service the company is constantly studying all causes of equipment and human failure and is designing the system so that service will not be interrupted when such failures occur. Interruptions are frequently caused by elements beyond the control of management, such as lightning, storms, and floods. Here efforts are made to restore service as promptly as possible. The state regulatory body is interested in seeing to it that all reasonable efforts are made to keep the service continuous and to restore it as promptly as possible when interruptions occur. More will be said later on the reliability of bulk power supply. It is sufficient to say here that continuity of service has reached a very high point. A study was made of the customer-hours of interrupted service of companies in the country over a ten-year period. This study indicates that on an average the customer can expect service 99.98 percent of the time. The aim is constantly to improve this record.

In some states the state regulatory body also has authority to see to it that service is rendered within the limits of good voltage regulation. The regulatory body may also provide that a customer cannot use a device that will make any sudden large demand of power for a very short interval that may cause a flickering of the lights in the neighborhood.

Usually the state regulatory body has the authority to see to it that adequate capacity is planned for the future so as to meet the power demands of the area. Usually a company planning to build a plant or transmission line must apply to the state regulatory body for a certificate of convenience and necessity for the building of such plant or line. The commission will grant such a permit if it decides that the line or plant is indeed necessary in the public interest. Thus the commission can prevent a duplication of facilities that will inevitably be against the public interest. A commission may find that the territory served already has adequate capacity in service or being installed and that therefore a proposed plant is not necessary. In this case the commission will deny the certificate.

Preventing Unjust Discrimination. Rate making is not an exact science. It is not possible to make the rate for each customer exactly cover the cost of furnishing service to him at all times. There are many factors entering into rate making.[1] As a practical matter customers are grouped into classes such as residential, commercial, industrial, street lighting, other public authorities, and railroads. Rates are designed to meet the general characteristics of the particular class.

In determining the fair value of the property for rate-making purposes and a fair return on the capital invested, the commission rightfully takes into account the property as a whole. It is possible to prepare an estimate of the earnings by classes of service, but this must be a separate analysis, as it would be very expensive and not practical or meaningful to keep books by classes of service.[2] The cost analysis showing

---

[1] See Chapter 8, "The Nature of Electrical Loads" and Chapter 9, "Costs and Pricing."

[2] See Chapter 9, "Costs and Pricing."

the earnings by classes is a tool that can guide management in designing the forms of the rates, but these analyses are really estimates prepared by the experts. The return by classes will vary from year to year as the load characteristics change from year to year.

Two of the principal factors taken into account in rate making are (1) the cost of furnishing service and (2) competition.

There are two principal kinds of costs entering into the cost of furnishing service: (1) the average cost and (2) the incremental costs. The analyst will use both of these kinds of costs in making the cost studies. As has been noted, the evolutionary process of building larger and larger generating units with higher and higher efficiencies results in a lower unit cost of making energy in the newer stations. However, it is the general principle that all customers should pay a fair share of all the costs of doing business if unfair discrimination is to be avoided. In other words, this means that all customers should share proportionately in the benefits of all the newer and more efficient stations.

At times a large industrial customer may feel that it should obtain all its electricity from the power plant nearest its plant site. Another may feel it would like to get all its energy from the power company's newest and most efficient plant. However, to avoid undue discrimination the commission requires that property as a whole be taken into account in fixing rates to the customers.

The principle of lowering rates as a means of increasing use of service is practiced generally in the power business. However, analysts and the commission are careful to avoid making any rate lower than the cost of furnishing service and especially lower than the full incremental cost of furnishing service. To so design such a rate would result in a decrease in percent return for all increased sales at the rate and result in discrimination against the ratepayers.

Meeting competition is another important factor in rate making. As long as a rate is less than a customer can provide for himself, he benefits through purchase from a company. It may be that some very large user can produce energy at a very low price. In such cases the company is justified in considering some incremental costs in designing the rate if the regulatory body is convinced that the business is not served at a loss to the company and at a penalty to other users. As long as the rate is above incremental costs, all other customers benefit through the taking on of a larger user. This is especially so if the user helps build load factor.

Incremental costs are generally considered those costs which are added to the company's over-all costs as business expands. They include the incremental cost of adding power plants and transmission facilities plus the added costs of generation. There may be other minor incremental costs, but these are the main ones.

Electric space heating is a kind of electric business that requires a rate to meet the competition of other fuels. As long as these rates stay above the incremental costs, all customers can benefit by the company's taking on this business. The electric space heating business is especially valuable in those areas where a company is experiencing a summer air conditioning peak.

Fair Return. So that the companies can keep on providing good and adequate service, the regulatory commission will generally approve rates which will permit them to earn a fair and reasonable return on the value of the property as determined by the commission. The commission cannot and does not guarantee that the companies will earn a set rate of return. Indeed, all during the years of the Great Depression of the 1930s most companies earned a below-standard rate of return. Some were quite low. The commission did not order rates raised to produce what might be considered a fair return. To have raised the rates in those years might have caused a decrease in revenue. As a matter of fact, during those years of low return the companies which were considered under good business management actually reduced rates further with commission approval in the attempt to build revenue through sales. There was great excess capacity, and little money was being raised to add to plant. The existing investors had to be content with what management could earn under those unusual conditions.

Many railroads today may be considered to be earning less than a fair return on the fair value of their property. But they are meeting severe competition from other transportation.

The idea that utilities are guaranteed a rate of return is a myth. The only guarantee is that there will be no unusual excess return.

While business is going through a healthy increase as it is today, and as people want and need more and more electricity, it is in the interest of the public as well as the investor that the commission allow the utility at least sufficient earnings to attract the capital in the free market required to build the plants to generate the additional kilowatt-hours. In addition some commissions take into account some factor for good management and allow the company sufficient return to do the research and provide the innovation that will enable the continued improvement in the efficiency of operation so as to continue the efficient operation and low price.

The Free Market

Under the American free-enterprise system, people or organizations are free to invest or not to invest in securities of companies such as electric utility companies. When they do invest, they become part owners of the company.

When a person wants to invest in securities or buy some shares of stock, he calls upon a broker. The broker acts as his agent in the buying and selling of securities. This trading, or buying and selling, of securities usually takes place at one of the stock exchanges located in the major cities. The largest one is in Wall Street in New York City. There are brokers in most cities who can be reached by phone or otherwise from all parts of the country. The person wishing to purchase a security tells his broker the price he would like to pay. The broker then is on the lookout for someone wanting to sell the securities desired by the buyer, at the price the buyer desires to pay.

The broker may find a seller, but they may not be able to agree on price. In that case, the broker will try to get these two parties together in agreement on price and arrange for the transaction and for the transfer of the securities. The price at which the sale is made is recorded at the stock exchange and becomes part of the daily quotations, which generally establish the market price of securities. These security prices are quoted every day in the principal newspapers for all of the securities traded. There are price quotations for bonds as well as for stocks. This is the free market.

Electric utility companies follow this market carefully so that they can take advantage of favorable market conditions. By doing this, they are able to borrow money at the most favorable interest rates and sell stock on a favorable basis.

The regulatory commissions also follow market prices and yields on utility securities. The reason for their interest is that the cost of money is one of the major costs of doing business in the power industry. This cost is usually considered in deciding what is a fair return to the electric utility.

As these electric companies are owned by investors, the electric companies are sometimes referred to as "investor-owned" utilities. As a whole, these companies are often called the "investor-owned electric utility industry." However, the word "company" itself signifies ownership and operation under the free-enterprise system. The words "electric light and power companies" or "power companies" or "electric utility companies" are often used in speaking of that portion of the industry financed in this fashion. These companies serve almost 80 percent of the nation's electricity customers.

The rest of the electricity customers in the country are supplied by government power projects or cooperatives. There are several kinds.

Federal Power Projects

Federal power projects are projects which have been financed and built by the Federal government. In some cases, the financing is through direct appropriations by the Congress, and in other cases, the capital is lent by the Federal government. The Tennessee Valley Authority (TVA) is an example of a project which in the past has been financed by appropriations from the Congress and by investment of earnings. As of 1966 TVA was authorized to sell $1,750 million in revenue bonds to the public, the proceeds to be used for future expansion.

Public Utility Districts

The laws of some states permit the organization of public utility districts. These are political subdivisions of the state. This type of organization is prevalent in the states of Washington and Nebraska. In Washington the district usually encompasses a county. In Nebraska the size may vary from that of a single city or town to a county or even a larger area.

Municipal Ownership

Since 1882, more than 4,300 municipalities have at one time or another established their own electric utility systems. By the end of 1965 only 1,959 of these were still in operation. The number has gone down mainly because most municipal plants are limited in size; usually each plant serves only the town or city in which it operates. Because of their small size, most of these municipally owned plants cannot make electricity as cheaply as the large interconnected systems of the power companies. Also it has been found that the small municipal power plants often do not have the power capacity necessary to serve large industries.

The municipal power plants are financed through the sale either of bonds which are the general obligation of the community or of bonds supported by the revenue of the municipal power system.

REA Cooperatives

In 1935 the Federal government began a rural electrification program designed to hasten the extension of electricity to America's farms. In 1936 the Rural Electrification Administration (REA) was created. The purpose of this program was to make Federal money available at low interest rates in order to encourage the extension of service into thinly populated rural areas. The Federal government lends money at 2 percent interest with a condition that the loan be repaid within thirty-five years. At the end of 1967, 925 rural electric cooperatives were active borrowers from the Rural Electrification Administration.

As of June 30, 1968, 98.4 percent of all of the farms in the United States had central-station electric service available. Of these farms, about 43 percent are served by the electric light and power companies, 51 percent by the rural electric cooperatives, and 6 percent by other suppliers such as public utility districts and municipal plants. The cooperatives bought about 37 percent of their total power requirements from electric utility companies during fiscal year 1967. The remainder was generated by cooperatives or bought from some government power project or system.

Generation by Type of Supplier

In 1967, America's electricity was generated by the various types of suppliers in the percentages shown in Table 4.1.

TABLE 4.1 Proportion of Electricity Generated by Various Suppliers (Percent)

Electric power companies	76.6%
Federal projects	13.4
Municipal systems	4.7
Public utility districts	4.3
Cooperatives	1.0

Valuations

The regulatory bodies of the nation have an intricate and complicated task to perform in looking after the interest of the public—as the consuming public and as the investing public. Generally speaking they are well staffed, and they carry out their functions with skill. The most exacting task is to find the combination of fair value of property and fair return on that property. There are people who have devoted their lives to these matters. Many valuable books have been written on this subject. There are university scholars who testify on these subjects and express their views in textbooks. The state regulatory bodies have at their command this wealth of knowledge and experience as well as the experience and judgment of other regulatory bodies meeting similar problems. Added to this the commissions have as their guide a great wealth of court decisions. The rate case when decided can be appealed to the lower courts by either party. These lower court decisions frequently are appealed to a higher court. Some cases reach the Supreme Court.

The lay student and public may find it difficult to judge the efficiencies of these regulatory bodies without intimate knowledge of the intricate procedures in each case. Possibly the best measure of the efficiency of the regulatory procedure is that the electric power industry has met all demands of the American people for electric service in times of war and peace and that the average price of electricity has continued downward despite the inflation which has caused price increases in most other commodities.

The best over-all measure of operating results of the Edison Power Company is the percent return on the total value of the property used in serving the public. The "amount available for return" is the amount left after all operating expenses and all taxes and depreciation are paid. This money is used to pay bond interest and preferred stock and common stock dividends; whatever may be left is retained in the business and reinvested. For the Edison Power Company, the amount available for return in 1968 was $39,886,194.

When the Edison Power Company is called before the commission to answer questions as to why the rates should not be reduced, or when the company finds it necessary to go to the commission for a rate increase, the commission sets a fair and proper value on the company's property for the purpose of determining a fair rate for electric service. In determining this value the commission takes many factors into account. It may consider the value as recorded on the books, the value of securities outstanding, the amount set aside for depreciation, and in some cases the present value of the property, or the cost of building the same plant today.

Various Considerations of Values. Here are some of the aspects of "value" that the commissioners may consider.

Original Cost. Edison Power Company, like most electric utility companies, keeps its books in accordance with the uniform system of accounts established by the Federal Power Commission. These records give the original cost of the company's facilities.

Most power companies, including Edison Power Company, also keep what is called a continuing property record of all of the items, large and small, that go into the making of the electric utility system and the year in which these various items were installed. Edison Power Company's electric system has been built piecemeal over the years, with additions and betterments made to meet the growing demands of its customers. The continuing property record shows, for example, that in 1935 the company installed so many poles of a certain classification, so many transformers of a certain size, so many pieces of hardware properly classified, so many pounds of wire of a certain size, and many other items. It also includes a record of power plants installed, with each plant broken down into its appropriate parts. For the year 1936 there is a similar record, and so on for each year. The cost as shown in this record is the original cost of the property. Of course, as a result of price rises since the equipment was installed, the cost of replacing the item in a current year may be considerably higher. If prices had gone down, it could well be that the reproduction cost of the property might be less than the total value shown on this record. Present-day costs of building electric utility systems are generally greatly in excess of original cost.

Net Original Cost. The depreciation reserve reflects the charges to current income estimated to be required in order to amortize the cost of the plant and equipment by annual amounts over the expected useful life of the equipment. When a unit of property is retired from service, the amortized amount less net salvage is then charged to the depreciation reserve. The annual charges to income permit the company to recoup the cost of the equipment over its useful life.

The amount of depreciation is an accounting entry based on the best judgment of power company experts and the commissions. Edison Power's engineers study the life histories of various kinds of equipment. Certain items may be expected to last twenty years before wearing out. Others may be expected to last thirty years. With these studies as a guide, management records certain annual amounts in the depreciation reserve.

A number of factors contribute to a decision to replace a piece of equipment, including physical wear and tear, inadequacy, economic obsolescence, needs of public authorities, and accidents. Obsolescence is an important factor in deciding to replace equipment. For example, new boiler-plant equipment and generating equipment which are more efficient than that which had previously been available have been, and are being, designed. There comes a time in the economic life of a power plant unit when the company would do best to take it out of use and replace it with a unit that can make electricity more cheaply.

As replacements are not made every year for most items, the depreciation reserve normally builds up over a period of years. For Edison Power Company, the depreciation reserve on December 31, 1968, was $178.6 million, which was 23.4 percent of the gross plant account.

Many regulatory bodies deduct the depreciation reserve from the total original cost of the utility to arrive at what is called the *net original cost* of the property.

In some cases, it is argued that the net original cost more nearly represents the value of the property, for the reason that the company did not have to borrow the money represented by the depreciation reserve in the free market, and therefore it does not have to earn a return on this property. When the depreciation reserve is deducted from the total original cost of the property, the result is that the company earns no return on that portion of the property built with cash from the depreciation reserve.

Others argue that the depreciation is a prepayment for the replacement of property when it wears out or becomes obsolete, both of which are inevitable. Management could put this cash in a savings bank and earn interest on it, but that would likely be something less than a return on utility property as the return takes into account some risk. If the reserve is placed in a savings bank, the interest of course would be credited to net income. The new money required to build the property would be all from the market, and the company would be allowed a full return on this just like any other property. Those who hold this view argue that the company obligates itself to furnish continuous and adequate service to customers receiving service from property built with the depreciation reserve cash just as they do on all other parts of the business. The company must maintain the property and keep it in good operating condition.

Value of the Securities. Edison Power Company, like all power companies, has a record of the book value of all of its securities outstanding. The commission may give some consideration to this value in finding the appropriate value for rate-making purposes.

Reproduction Cost. In the earlier years, many commissions relied upon a physical inventory and valuation of the property in determining its value for rate-making purposes. To make this valuation an appraiser would be hired to make a detailed inventory of the whole property, counting and listing all items of equipment by size and classification. Then the cost of building everything at current prices was figured. This was referred to as the *reproduction cost new* of the property.

Competent engineers familiar with electric utility property are also employed to determine the actual existing depreciation in the utility's property. This is done by an actual physical inspection of the property to determine loss in value from wear, tear, and action of the elements, and from accidents or storms. In addition the engineer determines the actual existing depreciation from obsolescence, inadequacy, public requirements, inefficiency in operation, and all other causes of depreciation. The loss in value from the total of these items measures the actual existing depreciation in property. It is deducted from reproduction cost new to determine reproduction cost new less depreciation.

It must be remembered that not all items of depreciation are cumulative. Frequently obsolescence will cause the replacement of property in excellent physical condition.

In finding fair value under the reproduction cost theory it is generally considered proper to deduct the observed depreciation in order to find the present depreciated value of the property.

Trended Present Cost. As most companies now have fairly complete property records, a close approximation to this reproduction cost can be found by what is called the trending process. Indexes such as the Handy-Whitman Index are used in finding the present value, working from the original cost figures. (The Handy-Whitman Index is a refined estimate by utility engineers and appraisers of the cost of basic items used by the utilities, by year. It is roughly equivalent to a "cost of livng" index for power companies.)

By way of example, the property records may show that the company purchased a certain number of 5-kilovolt-ampere transformers in 1955 at a certain price. This figure is recorded for that year. Using the Handy-Whitman Index for transformers, that cost can be converted into the current cost of those particular transformers. In similar fashion, all items of property can be trended forward to determine their present value. Since there is a record of what property is in service, it is not necessary to have engineers go over the property to count the items. The current value of the property, arrived at in this way, is referred to as the *trended value* of the property. It approximates the reproduction cost of the property.

Besides this trended value, the observed depreciation is sometimes determined by hired engineers, as is done in the case of reproduction cost. The trended value less the observed depreciation becomes then a close approximation to reproduction cost less observed depreciation.

A regulatory body may take into account all of these elements of value and then arrive at its own opinion as to what is the proper and fair value of the property to use as a base for the fixing of rates. Before finally determining which method or combination of methods it will use in determining fair value, the commission will hear experts on the cost of money. The commission may first be inclined to give most of the weight to net original cost as the rate base. But this value may result in insufficient net revenue to attract capital from the market. If this is so, the commission's views on proper rate base may be influenced in some respects by its views as to proper return.

Values for Rate Making. For many years the courts have held that the value of the property and equipment used was the basis for determining the allowable return. Since the commissions must follow the rules laid down by the courts, there are many commission decisions which say value is the proper basis of return.

Frequently courts and commissions, instead of using the term "value," have held that utility companies are entitled to earn a return on the "fair value" of the property devoted to public service. In the leading case on this subject (Smyth *v.* Ames), decided in 1898, the Supreme Court of the United States said:

> The basis of all calculations as to the reasonableness of rates to be charged by a corporation maintaining a highway under legislative sanction must be the fair value of the property being used by it for the convenience of the public. And in order to ascertain that value, the original cost of construction, the amount expended in permanent improvements, the amount and the market value of its bonds and stock, the present as compared with the original cost of construction, the probable earning capacity of the property under particular rates prescribed by statute, and the sum required to meet operating expenses, are all matters for consideration, and are to be given such weight as may be just and right in each case. We do not say that there may not be other matters to be regarded in estimating the value of the property. What the company is entitled to ask is a fair return upon the value of that which it employs for the public convenience.

In the Natural Gas Pipeline case in 1942 the Supreme Court seemed to abandon its former position in the field of public utility regulation. Upholding a rate-fixing order of the Federal Power Commission, the Court said:

> The Constitution does not bind rate-making bodies to the service of any single formula or combination of formulas. Agencies to whom this legislative power has been delegated are free, within the ambit of their statutory authority, to make the pragmatic adjustments which may be called for by particular circumstances. Once a fair hearing has been given, proper findings made, and other statutory requirements satisfied, the Courts cannot intervene in the absence of a clear showing that the limits of due process have been overstepped. If the Commission's order, as applied to the facts before it and viewed in its entirety, produces no arbitrary result, our inquiry is at an end.

In 1944 in the Hope Natural Gas Company case, the Supreme Court swung even further from its former position when it said: "Under the statutory standard of 'just and reasonable' it is the result reached, not the method employed, which is controlling. It is not theory but the impact of the rate order which counts."

Following these two Supreme Court decisions, a large number of commissions adopted "original cost" or "prudent investment" as the preferred method of valuation. In 1948 the Federal Power Commission reported eighteen state commissions using one or the other of these methods. In 1966 the number was twenty-six.

During the past few years there has been evidence of a trend back to the fair value concept. In February, 1956, the New York Court of Appeals ruled that the "commission is required to receive proof of reproduction cost less depreciation as some evidence of present value in the case of utility property."

(In 1945, about one year after the Hope decision, the New York Public Service Commission adopted the original cost method of valuation.)

In 1956 the Mississippi Legislature passed a law enlarging the authority of the Public Service Commission to include gas and electric utilities. The new law specifically provides that the "rates prescribed by the commission shall be such as to yield a fair rate of return to the utility furnishing service, upon the reasonable value of the property of the utility used or useful in furnishing service." Previous to the passage of this law, Mississippi had been an original-cost state. Recent commission and court rulings indicate, however, that it may still be so—at least for the present.

Many scholars have different views as to the proper values for rate-making purposes, and these views vary among the states. There is no pat formula for finding the answer to these complicated matters. The process requires wisdom and judgment on the part of the regulatory bodies which analyze each case on its merits after hearing all the appropriate facts and opinions on each case.

Return. Having found the value of the property for rate-making purposes, the commission then usually sets out to find the proper or fair return to be allowed on that value. Commissions try to find a return which is fair to existing investors, will enable the company to attract the required new capital in the free market, and will enable the company to maintain a good credit position.

The commission will weigh the state of the stock market, which affects the current cost of money. Very likely an expert in the market will be asked his opinion as to what is a fair return. He might also be asked what the minimum return should be in order to enable the company to raise money in the free market for expansion to meet the customers' demands.

After weighing all these factors, the commission reaches a decision, states the value of the property for rate-making purposes, and sets a certain percentage as a return to be allowed on that value. This percentage multiplied by the valuation results in the earnings or return which the commission has allowed.

Having thus set a figure for these earnings, the commission

then orders an adjustment in rates which will produce these earnings. If the earnings set by the commission are higher than the company is currently making, a rate increase is in order and the company is asked to prepare new rate schedules which will produce the higher revenue. If the earnings allowed are less than current earnings, the company is asked to file new rate schedules lower than those currently in effect.

During the 1930s and early 1940s the Edison Power Company, like most of the electric utility industry, was reducing its rates on a voluntary basis. During World War II the company was unable to buy new equipment, and in the period following the war it was hit by inflation. Consequently it could not realize the normal benefits of economy from installing new machines, and a rate increase was inevitable. As the company began to feel the effect of the new larger and more efficient generators, of increased sales, and the greater degree of interconnection and coordination it was able to develop, it was able to resume the pattern of rate reductions. Now, with unusual increases in cost of capital and labor, and with increased taxes, the company is considering a rate increase.

Regulatory Procedure

The Edison Power Company continually files with its regulatory body all its appropriate financial statements and operating reports. Thus the regulatory body has a continuing knowledge of all the company's operations and procedures. The commission knows about proposed plans for construction, for it must issue appropriate certificates showing their convenience and necessity. At any time the commission can, and does, ask for pertinent information and receives it.

Any time the company feels a need for a change in any pricing schedule, it must have commission approval and show justification for making the change. The change requested may be for a reduction in price, in which case a formal proceeding may not be necessary. If the commission feels a rate reduction of substantial amount is in order and the company questions the advisability of the amount of the reduction, a formal rate proceeding might ensue. If the company feels that a rate increase is in order a formal proceeding would be required.

In such a formal proceeding the company presents its whole case in testimony before the commission. Management people and appropriate experts testify. The company may employ outside experts to give independent judgment as to value of property, justification for the proposed change, and rate of return. The witnesses may be cross-examined by the commission counsel with the advice of the commission staff. Interested parties may intervene, put in testimony, introduce expert witnesses, and cross-examine the company's witnesses. Often the proceedings go on for weeks and months while the commission weighs all the evidence and issues an order giving its findings and conclusions. These may be accepted by the parties concerned or reviewed by the courts.

All the company's records of revenue and expenses are open to inspection by the commission. The expenses are reviewed by the commission as a matter of regular routine and specifically inspected in rate cases.

State laws vary and the practices of state regulatory bodies vary in the treatment of some of these expenses. Among the items frequently debated are the treatment of taxes in connection with (1) accelerated amortization, (2) liberalized depreciation, (3) investment tax credit, and (4) guideline depreciation.

Accelerated Amortization and Liberalized Depreciation. In 1940, Congress incorporated into the Internal Revenue Code of 1939 an amortization provision permitting corporations in calculating taxable income to amortize over a five-year period such portion of investments in emergency facilities certified by the government as necessary in the interest of national defense. The purpose of the provision was to eliminate uncertainty and the inability under the then existing law to predetermine the rate of depreciation on expenditures requested by the government in the expansion necessary to produce material for national defense.

This provision was carried forward in the Internal Revenue Code of 1954. Also in the Internal Revenue Code of 1954 provision was made for permitting corporations to accelerate the depreciation taken for tax purposes on new construction over the normal straight line depreciation to the end that about two-thirds of the cost of an asset can be written off during the first half of the property's life. It was the intent of Congress that this provision would have far-reaching economic effects in providing the incentives vital in helping to create new jobs and to maintain a high level of investment in plant and equipment. By allowing businesses to recover a large part of their costs more quickly, the financing of expansion is aided. This was part of Congress's desire to maintain a virile and growing economy in face of the cold war with the Soviet Union.

As in the case of accelerated amortization, not more than 100 percent of the cost of a facility can be depreciated for tax purposes; so if two-thirds of the cost is depreciated in the first half of the property's life, only one-third remains to be written off during the second half. Consequently, any tax deferral received during the first half is fully made up in the second half (providing, of course, the facility continues to be income producing), and no revenue is lost to the government over the life of the property.

The net effect of both provisions is to defer taxes to some future date—not to reduce them over the life of the property.

During World War II and the Korean War the electric utility companies made capital expenditures for which certificates were granted. Also, a majority of the companies have taken advantage of the depreciation provision in the Internal Revenue Code of 1954.

In general, the handling of taxes deferred due to accelerated amortization and liberalized depreciation has been by one of two methods. Since this is a tax deferral and not a tax reduction, some companies (and some state commissions) treat the amount of tax deferred as a charge to income in lieu of a tax payment with a credit to a reserve for the payment of future income taxes. On the expiration of the tax deferral (five years in the case of accelerated amortization and approximately one-half of the life of the property in the case of liberalized depreciation, depending on the method employed) income is credited with the amount of increased taxes due to a lower depreciation credit and the reserve charged with such amount. This method "normalizes" earnings over the life of the facility. In some states such reserves are deducted from the rate base; some state commissions take it into consideration in determining rate of return; and other commissions treat it as a normalizing operation which then has no effect on charges to customers nor income accruing to the stockholders. Cash is generated during the period of the tax deferral and reimbursed during the remaining life period.

Other companies and commissions disregard the tax deferral as a charge to income and permit it to "flow through" to surplus as a part of earnings, on the assumption that taxes deferred are a regularly recurring item and subsequent deferrals will offset subsequent increased taxes. To the extent that deferrals exceed reimbursements, they are ultimately reflected in charges to customers through either rate reductions or the forestalling of rate increases.

From 1954 through 1960, those companies that have normalized deferred taxes have charged a total of $1,214,000,000 to income. During this period their total Federal income tax liability was $7,570,000,000. Deferred taxes amounted to 16 percent of their total Federal income tax liability. Capital expenditures of the investor-owned electric utility companies during this period amounted to $22,600,000,000. The cash generated by normalizing deferred taxes amounted to only 5.4 percent of such expenditures.

Accelerated amortization and liberalized depreciation have been available to all taxpayers having property used in a trade or business, or property held for the production of income. They have been available to government power projects, municipal plants and cooperatives financed by the Rural Electrification Administration, as well as to the investor-owned electric utility companies. However, government-owned and -financed entities pay no Federal income taxes, so they cannot, and, of course, have no reason to, take advantage of any deferral of taxes afforded by the Internal Revenue Code. They are already getting a 100 percent tax exemption.

Company Contributions

The Edison Power Company, like all public utilities, is called upon to make many contributions in support of community affairs. In fact, it has become commonly accepted practice for corporations of all kinds, as well as individuals, to make contributions in support of various community activities. The Edison Power Company is asked to help support the Community Chest, the hospitals, the YMCA and Boy Scouts, colleges, educational foundations, and the like. To refuse to contribute would mark the company as lacking in community interest, and this in turn could adversely affect its sales and revenue. Such contributions are generally accepted by regulatory bodies and the Internal Revenue Service as legitimate business expenses. However, in each case the state regulatory body has the opportunity to pass on these items and to decide whether the charges should be placed "above the line" and thus allowed as an operating expense or charged "below the line" and thus not allowed as an operating expense. In each case the company is notified whether the particular contribution will be allowed as a deductible item by the Internal Revenue Service for income tax purposes.

Wages and Salaries

The company's total wage expense is filed with the commission along with other expenses as part of a standard classification of accounts. The salaries and wages paid by the company are considered a matter of management decision. The company must pay the going wages in the community for the type of employees it needs or else it is unable to employ people and to give good service. There is a market value for labor in all categories, just as there is a market value for poles and wire. The Edison Power Company tries to employ the most competent management people it can, because their skills result in good economical management, innovation, and new developments—all of which are of value to the consuming public and to investors.

Expanded Regulatory Powers of the Federal Power Commission

As has been mentioned, the Federal Power Act gave the Federal Power Commission authority over transactions where energy is sold at wholesale in interstate commerce. In this case, "wholesale" means the sale of energy to an organization or agency which in turn retails the energy to a consumer. Energy is said to be in "interstate commerce" when it is generated in one state and delivered across state lines for use in another state. Up until 1964 it was generally understood that the Federal Power Commission had jurisdiction over those transactions where the energy moved directly from company A in one state to company B in the adjoining state. The energy entering the system of company B is commingled with all of the energy generated by company B and becomes a part of company B's over-all power supply. Company B sells this energy to its customers and may also sell energy at wholesale to small investor-owned companies or municipally owned electric systems which in turn retail the energy to the customer.   All these transactions, including the wholesale transactions of company B, were considered to be intrastate matters and under the jurisdiction of the state regulatory body.   It was understood that the Federal Power Act was designed to regulate those matters not being regulated by state regulatory bodies.

There may have been cases when a company did not have sufficient generating capacity to meet the requirements of a customer buying energy at wholesale, in which case some of the energy came from outside the state and flowed in interstate commerce.   In such an instance the Federal Power Commission had jurisdiction, but if it could not be shown that the energy delivered to the customer at wholesale came from outside the state, the state regulatory body exercised jurisdiction.   In a number of cases the Federal Power Commission asked that these wholesale rates be filed with the FPC with the understanding that they were simply for the information of the Commission and did not imply FPC jurisdiction.

In considering a proper price for energy sold at wholesale, the state commission considered most of these wholesale buyers as it would any other purchasers of energy in similar amounts and under similar conditions.   That is, wholesale buyers of energy were treated like any other class of customers.   Most of them were small municipal systems.   The kilowatts of demand, the kilowatt-hour use, and load factor of these systems were similar to those of large lighting and power customers such as department stores, hotels, shopping centers, or large office buildings.   In some states the commission classified all this service in the same category and gave the purchasers the same rate.   In other states there was a class rate for wholesale to municipalities, but the level of the rate was kept in line with the rates of other customers buying similar amounts of energy under similar conditions.   In this way the commission avoided unjust discrimination.   Municipal power systems can and frequently do generate their own energy, but the number of municipal systems generating electricity dropped from about 831 in 1940 to 392 in 1964. The systems which shifted to purchasing energy at wholesale found they could get better service at a lower price by getting the energy from the large, more efficient interconnected power systems. In fixing the price schedule for these large lighting and power customers, both at retail and wholesale, state commissions seldom enter into a rate case for this class alone, any more than they would go into a rate case for residential customers as a class or commercial customers as a class. If there is a reasonable balance in earnings among the various classes the commission considers the property as a whole in determining whether the company is earning a fair return. If an over-all increase is in order, the combined judgment of the company management and the commission would indicate how to spread the increased cost of production over the various classes, always keeping in mind the cost of furnishing service and the need for meeting competition. This system of regulating the wholesale rates to municipal plants seemed to work well. Under it, the customer is always free to make its own power or, if it chooses, to buy power with the commission's assurance that it will be treated the same as any other customer buying service in like amounts under similar conditions. As the property as a whole is taken into account in fixing rates, all customers get their proportionate share of the less efficient and more efficient generating stations.

The Colton decision of 1964 changed this procedure. Energy for the City of Colton in California was being sold at wholesale by Southern California Edison Co. The City of Colton had a demand of about 9,000 kilowatts, whereas Southern California Edison had a capacity of about 6,000,000 kilowatts. Some energy in relatively small amounts flowed into Southern California Edison System from outside the state. The Federal Power Commission argued that when energy coming from outside the state, no matter how small the amount, enters into a power system, some of that energy must inevitably find its way into each wholesale transaction. The state commission and the company argued otherwise. They pointed out that in any event the state commission was exercising jurisdiction, that it had been for years, and that the Federal Power Act was intended to supplement and not replace state regulation. The Supreme Court finally decided the case in favor of the Federal Power Commission. As a result of this decision all wholesale transactions which had formerly been under the jurisdiction of the state commissions came under the jurisdiction of the Federal Power Commission, where the Colton principle applied.

Sharing jurisdiction over transactions within a state by the Federal regulatory body and the state regulatory body inevitably presents problems. The Federal Power Commission and the state commission may have different opinions as to a fair return and the fair value of property for rate-making purposes. The state regulatory body must of necessity take into account the earnings of the property as a whole and not any particular class of service in determining whether earnings are fair. The Federal Power Commission is interested only in one class of service—wholesale. If the Federal Power Commission fixes this rate, how will the rate compare with those of other customers buying service under similar conditions in similar amounts which have the rates fixed by the state regulatory body? Will the Federal Power Commission consider the earnings of the company as a whole before fixing the rate to the one class under its jurisdiction? Will the Federal Power Commission take into account the discrimination that might result if customers under its jurisdiction are allowed to buy energy at a price that is lower or higher than the rates of other customers buying similar service under similar conditions? All classes of service do not earn precisely the same percent return. Some may be below average and some above, and these change with changing load patterns.

Will this result in duplicating regulation? A rate case is expensive. Must a company go through two complete rate cases and thus double the expense of regulation, which expense is inevitably reflected in the ultimate price that the customer pays for the service? These are questions that have not yet been fully answered.

A case in point arises in connection with the construction of the large atomic power plants now being installed by investor-owned companies in a number of parts of the nation. It is claimed by some that small municipal plants cannot afford to install large atomic power plants themselves and so should be allowed to own a small fraction of one of the atomic power plants being installed by the companies. Is this not the kind of unjust discrimination that regulatory commissions were set up to prevent? The company is able to install the atomic power plant because it has thousands of customers that have used large amounts of energy. It has built up a large interconnected power system that can justify the new atomic power plant. The price of service to these customers is based upon all the costs of the company, the imbedded costs as well as the costs of the new plant. All the customers are entitled to their share of the output of new power plants. The municipal plants should not get energy from the newest most efficient plant alone, but should share the energy of the whole system. Any customer would like to have its rate based upon the cost of the most efficient plant in the company's system, but this has been considered unjust discrimination. Each new plant the company builds is a little more efficient than existing plants. The new plant may use conventional fuels or nuclear fuel.

Since the Colton decision, the Federal Power Commission has the authority to fix the price for wholesale service to the municipal systems when a possibility exists that small amounts of energy flowing to the supplier from outside the state may find their way to the municipal systems. It is a somewhat new field for the Commission. It has not had to weigh all the factors of all the classes of customers with all the ramifications of rate making that are faced by the state regulatory body.   To what extent will the Federal Power Commission give consideration to the views of the state regulatory body?   Only time will tell.

The System Works

The record shows that, in general, the regulatory system works well and fairly for the customer.   It is a system under which businessmen run the business and men in government act as the regulators to maintain honesty and fair play.

From a capacity and production standpoint, America has far more power than any nation on earth.   With only 6 percent of the world's population, Americans use over a third of the world's electric power output.   This is more than the next five countries combined.

Russia's social, political, and economic system, the exact opposite of all that America stands for, has often been held up as a challenge to the American way of life.   Russia has only 15 percent of the world's electric power, while its population is 18 percent greater than that of the United States. The electric energy used by Americans is almost three times the energy used by the Russians.

Chart 4.3 shows that this system has met all demands for power in both war and peace.   Electricity is one of the few commodities and services that met the demand during the last great war.   From the standpoint of power capacity and supply, the American system of operation by citizens and regulation by government has worked well.

From the pricing standpoint, the record also is good. Chart 4.4 shows that the average price of a kilowatt-hour of residential electricity today is only 25 percent of what it was in 1913.   The cost of living today is now three and one-half times what it was in 1913.

Part of the decrease in the average price is occasioned by the fact that the rate schedule provides for an automatic decrease in average rate for all increased use.   The average use

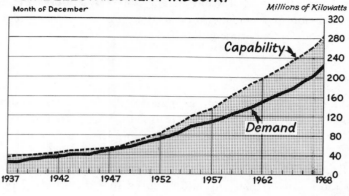

CAPABILITY vs DEMAND
TOTAL ELECTRIC UTILITY INDUSTRY

Month of December   Millions of Kilowatts

CHART 4.3

Electric Utility Industry PRICE RECORD

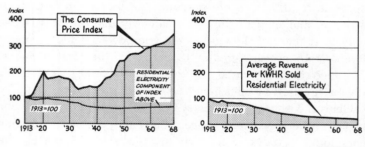

CHART 4.4

has been constantly increasing. Despite the great increases in the costs of material, labor, supplies, and taxes, the electric utility companies under government regulation have been able to maintain low prices and to reduce the average cost.

With the widespread direct and indirect ownership of the investor-owned electric companies, it would be hard to find one family in the United States which was not in some way economically and socially concerned with the progress of the electric utility industry and which has not benefited from it.

5

Economic Trends and Their Effect on the Power Company

The economic factors having the greatest effect on the electric utility business are, in general, not short-range ups and downs, but those which exert their influence over a long period of time. New construction is planned to handle five or six years' anticipated growth. The effect of decisions made today may not be seen for years. Trends in this industry have a way of continuing in the same general direction over the long run. These trends can be used to help the companys' officers make the right decisions.

Of course the electric utility business is affected to some extent by short-range business conditions. When business is good and employment is high, people have money to spend and sales of electricity rise with other sales. When general business conditions are not so good, people spend less and the growth rate of electricity sales may slow up. These short-range changes in business conditions affect the sale of electricity to industry more than they do sales to other kinds of customers. Residential service is usually the least affected by such changes; it is the most stable of all the classes of service. People may cut back the operation of their factories in a recession, but they generally run their homes according to their usual pattern.

Since the long-run trend in the use of electricity is upward, a recession which would cause a drop in sales of ordinary commodities may cause a mere leveling off of sales of electricity. The electric power industry is one of those least affected by a business recession.

Wholesale Price Index

The government keeps a record of the wholesale prices of most commodities and from this prepares a "Wholesale Price Index." This is one of our best barometers of business conditions. Chart 5.1 shows the index from 1801 through 1968.

CHART 5.1

WHOLESALE PRICE INDEX

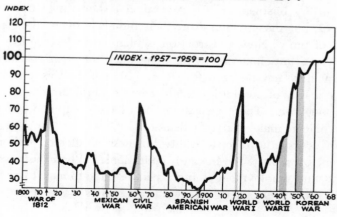

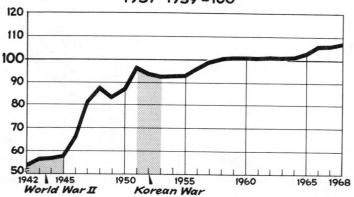

Wholesale PRICE INDEX
1957-1959 = 100

CHART 5.2

The major wars are indicated by shading. Note the sharp rise in prices before and during each major war and the sharp decline soon after the war is over.

A price decline started after World War II, but then came the Korean conflict and the general preparedness program. Today, prices of labor and commodities are still at a higher level. Because of this, the cost of most of the items entering into the cost of making electricity, like all other costs, continues at this higher level. This can be seen in greater detail in Chart 5.2.

No one knows what the future may bring. It is certain that government spending for defense has a lot to do with the high level of the index. There is every indication that the government plans a high level of spending over the next few years. So the index is likely to remain at a high level, for the near future at least.

Trends in Capability and Operations

Here are some of the trends of Edison Power Company. Chart 5.3 shows the kilowatts of demand and the kilowatts of

CAPABILITY *vs* DEMAND

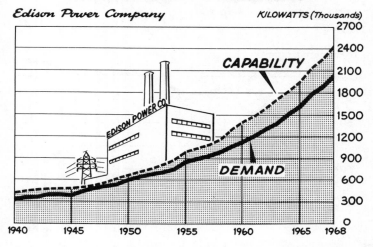

CHART 5.3

capability of the company from 1940 to 1968. Notice that the capability has remained above the demand throughout the whole period—in times of war as well as in peace.

Chart 5.4 shows the reserve generating capability of Edison Power Company as a percent of peak load. Notice that the

CHART 5.4

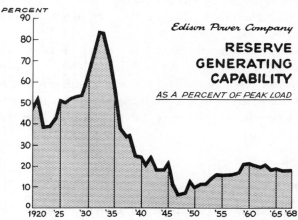

reserve rose to over 80 percent of peak load during the Depression of the early thirties. This means that little more than half the company's capability was called into service to meet peak load at that time. The reason for this is not so much that the company bought a lot of new generating equipment, but that many businesses had failed and factories cut back their operations. Because of this cutback, the demand for electricity was less and the company had many idle generators.

World War II brought with it a large demand for electric power to run defense plants. Because of wartime restrictions on the availability of generators, Edison Power's reserve capability hit a low of 6.1 percent after World War II. Since then the reserve margin has risen to 18.6 percent in 1968. (The margin of reserve for the entire electric utility industry in 1968 was 17.2 percent based on calendar peak.) Another reason for the decline after World War II was that people could then buy things that had not been on the market during the war. The company's customers were able to get new appliances, but the company had to wait three years before its new generating equipment was installed.

Load Factor. Load factor is one of the most important economic factors in the electric utility business. Chart 5.5 shows the annual load factor for the Edison Power Company from 1930 through 1968. For the most part there has been a gradual rise over the years, which is a healthy sign. Load factor dropped after World War II as industry went off the wartime round-the-clock operation and resumed its normal operation. There was a gradual rise from 1946 to 1957. Since 1958 load factor has been fairly steady (see Chapter 7 for the company's plans to improve load factor).

Since the early days of the electric utility industry, the predominant peak months have been in the winter, usually in December. The evenings are long and lighting is used for more hours of the day. Also industrial load is usually larger in the winter than in the summer.

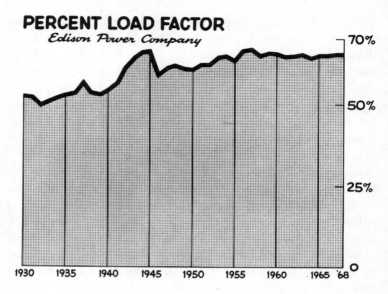

PERCENT LOAD FACTOR
Edison Power Company

70%

50%

25%

1930   1935   1940   1945   1950   1955   1960   1965 '68   O

CHART 5.5

In recent years there has been a shifting to a summer peak brought about by the rapid growth of air conditioning load. In Edison Power Company's service area the peak is still a winter peak. However, if it changes to a summer peak and the winter peak does not go up in proportion, load factor will go down.

There is a tendency for the load factor to increase as the summer peak approaches the winter peak. When the two are equal, load factor is higher. Then as one peak gets bigger than the other there is a decrease in annual load factor.

Power companies watched the experience of companies in the South and Southwest, where the load factor dropped as the summer peak increased over the winter peak. Many companies in other parts of the country have experienced a similar decline in load factor, some to a lesser extent, particularly where companies could promote the sale of electric house heating equipment to offset the air conditioning load. Chapter 7 shows how Edison Power Company is planning to meet this situation.

Chart 5.6 shows the trend in plant investment and gross annual revenue for the years 1930 through 1968.

Chart 5.7 shows the trend in gross operating revenue, taxes, operating expenses, depreciation, and balance for return.

Chart 5.8 shows, for Edison Power Company, the amount of investment required for $1 of annual sales. This is the *investment ratio,* the ratio of the plant investment to the gross operating revenue. Note the rise to almost $7 of investment per $1 of sales in 1933 when the company's sales were low because of the Depression. The investment ratio has come down substantially since then, but it is somewhat higher now than in 1947–1948 when the margin of reserve was at the low point.

Operating Trends by Periods. In analyzing these various trends, it appears that the company has been affected by dif-

CHART 5.6

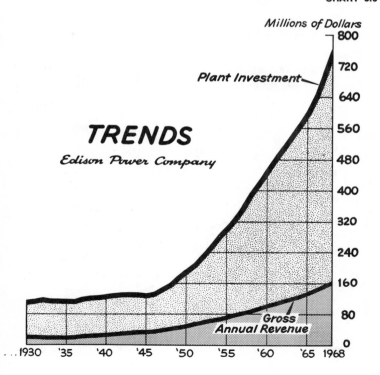

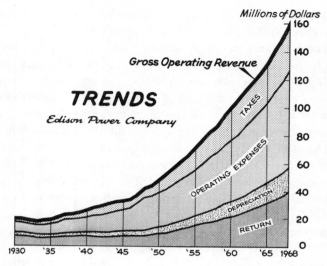

CHART 5.7

ferent circumstances during different periods of its growth.
At a recent staff meeting, Edison Power Company's economists and top management reviewed the company's growth
and its effect on these trends.   They found that the company

CHART 5.8

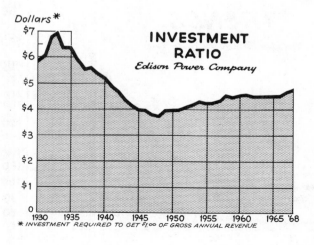

went through six significant periods, each of which produced marked changes in the nature of the company's operations.

Period I—1931 to 1933. These were the years of the severe slump in business caused by the Great Depression. Total sales, net income, and percent return on investment all dropped. Edison Power's management was doing all it could to cut expenses during this period, but despite these efforts it could not prevent a severe drop in percent return on the investment. Notice also from Chart 5.4 that the margin of reserve of Edison Power Company was abnormally high during this period. This was largely the result of an attempt to continue building new plants in order to lessen the effect of the Depression. This was done in response to President Hoover's request that capital expenditures be continued.

The peak load on the company dropped after 1930, with the result that it, like most other companies in the industry, had considerable excess capability during the thirties.

Period II—1934 to 1940. This is the period of the long climb out of the Depression. Sales of electricity began to rise as people had more money to spend and as companies stepped up their sales efforts. During this period a number of companies, including Edison Power, voluntarily lowered their rates with commission approval. They did this in the hope of increasing sales, even though earnings at the time did not justify rate reductions.

During the 1934 to 1940 period taxes, operating expenses, and depreciation also increased.

Largely because of the rate decreases during the period, Edison Power Company showed hardly any rise in net return during the period 1934 through 1940, even though sales had increased considerably. This was not overly important because plant investment did not go up much in this period. As a matter of fact, investment decreased slightly because the value of old equipment taken out of service during the period was greater than the amount of new construction.

Period III—1941 to 1945. About 1940, Edison Power

Company saw a coming need for more power capacity. Plant reserves were nearing 20 percent and dropping rapidly. As capacity was added, the investment in Edison Power showed a slight rise beginning about 1940. However, there was still no increase in net income. Taxes, operating expenses, and depreciation continued to go up. These rising costs offset any savings that could be brought about through increased efficiency. Taxes were beginning to make themselves felt in a substantial way.

The utility plant investment leveled off again beginning about 1943. The reason was that the company was unable to buy generators and other equipment; the manufacturers were busy building turbines and generators for the armed forces. The electric utility industry had to find ways to get along without new capacity and had to utilize all of its old and somewhat obsolete facilities. The increased use of the older equipment resulted in higher costs.

Period IV—1946 to 1960. This was the period of post-war expansion. Total sales rose sharply, taxes continued to increase, and other expenses went up in proportion.

Despite the continued rise in sales, the balance for return still remained almost level until about 1947 or 1948. As a result, from 1945 through about 1948 the percent return on investment declined.

An increase in net operating revenue began about 1948, resulting from the increased operating efficiencies of the new generators coming into operation and higher rates.

All through this period both sales and expenses went up. Net income also increased. Percent return held about the same from 1949 through 1955–1956. There was a slight decline in 1957.

It was during this period that a number of companies had to reverse their longtime practice of reducing rates and ask the commissions to grant some rate increases. They were feeling the effect of those years when they had been unable to purchase newer, more efficient machines, and the effect of the new machines they were then purchasing had not yet had sufficient impact on lowering production costs. However, despite these rate increases the average price of residential service continued its downward trend. This was because the sliding scale nature of electric rates includes a built-in feature that results in automatically lowering the average price with increased use.

Period V—1961 to 1967. This was a period of continued expansion. Generating unit sizes increased, and Edison Power Company began using transmission voltages of 345,000 volts and higher. Interties with neighboring systems were strengthened and pooling agreements tightened. The company began planning for a nuclear plant, and electric space heating began to be a meaningful part of the company's load.

Increased efficiency of operation and increased use by customers made it possible for rate reductions to be made during this period. However, costs were going up and, particularly in the latter years of the period, were having an effect on percent return.

Period VI—1968. The Edison Power Company views 1968 as a turning point. The evolutionary pattern of growth toward larger units, higher transmission voltages, and more interconnection will continue, but the tightness of the money market and the increase in costs of construction, taxes, and labor have created a squeeze. Management is concerned because in this situation it will be difficult, and in some cases impossible, to hold rates level.

Increased Costs Mean Greater Investment

Chart 5.9 shows the percent increase in the cost of some representative items used in generating electricity over the past ten years. Poles and fixtures are up 34 percent, overhead conductors and devices up 41 percent, underground conduits up 37 percent, and common labor is up 62 percent. These

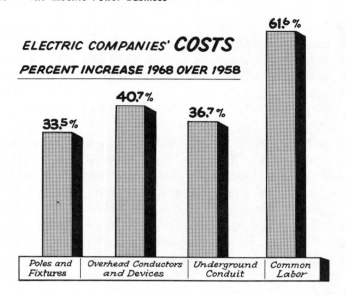

ELECTRIC COMPANIES' COSTS

PERCENT INCREASE 1968 OVER 1958

61.6%

40.7%

36.7%

33.5%

| Poles and Fixtures | Overhead Conductors and Devices | Underground Conduit | Common Labor |

CHART 5.9

and other higher costs have had their effect on the long-range trend in the operations of Edison Power Company, because they increase the amount of money the company must invest in equipment. The company has been able to bring about some savings which have helped somewhat in offsetting these rising costs.   Here are some of the ways the company has realized these savings.

Chart 5.10 shows the maximum size of generator units purchased by electric companies for 1920, 1928, 1963, and 1965. The larger machines are more efficient to operate and cost less per kilowatt to build than small generators.   Table 5.1 shows, in approximate figures, how this works out.

With the larger plants also came other savings in building space, piping, and so forth.

The electric utility companies have been using higher and higher transmission voltages, as shown in Chart 5.11.   It is expected that transmission line voltages as high as 750 kilovolts will come into use during the early 1970s.   Generally speak-

Maximum Size of Generator Units

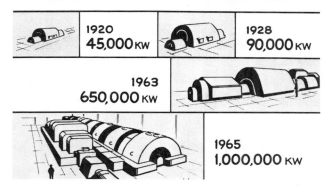

| 1920 | 1928 |
| 45,000 kw | 90,000 kw |

1963
650,000 kw

1965
1,000,000 kw

CHART 5.10

ing, the higher voltages result in a lower cost per kilowatt of carrying capacity.

In like manner the companies have increased distribution voltages so that the lines can carry more electricity. This has helped keep down the investment in distribution facilities.

TABLE 5.1   Size of Units and Cost

Size of generating unit (kilowatts)	Installed cost of unit (per kilowatt)	Approximate cost ° (mills per kilowatt-hour)
100,000	$148	6.0
200,000	$130	5.5
500,000	$115	5.1
1,000,000	$ 99	4.8

° Includes power plant fixed charges plus production expense.

But these economies have been offset by the general rise in the cost of labor and equipment. The result is shown in Chart 5.12. Notice that investments in transmission and distribution facilities have increased slightly in dollars per kilowatt. As a result, the total investment in the power system of Edison Power Company increased from about $348 per kilowatt in

Maximum Transmission Voltages

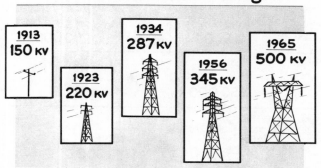

CHART 5.11

1958 to about $366 per kilowatt in 1968.   Owing to the savings mentioned above, this increase has not been nearly as great as the general rise in the cost index.

Chart 5.13 shows the rise in the Handy-Whitman Index (see page 89) of utility costs for production facilities.   Notice that from 1958 to 1968 the index has increased from about 747

CHART 5.12

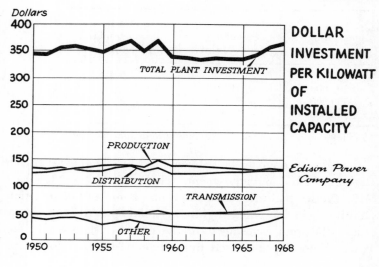

DOLLAR INVESTMENT PER KILOWATT OF INSTALLED CAPACITY

Edison Power Company

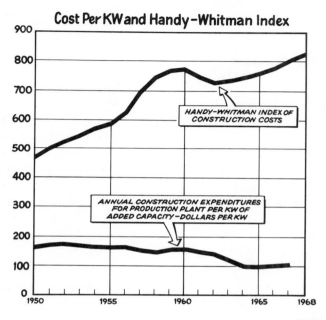

Cost Per KW and Handy-Whitman Index

900 — 800 — 700 — 600 — 500 — 400 — 300 — 200 — 100 — 0

HANDY-WHITMAN INDEX OF CONSTRUCTION COSTS

ANNUAL CONSTRUCTION EXPENDITURES FOR PRODUCTION PLANT PER KW OF ADDED CAPACITY—DOLLARS PER KW

1950   1955   1960   1965   1968

CHART 5.13

to 830.   If it were not for the economies brought about by larger units, higher voltages, improvements in manufacturers' efficiency, and other factors, the unit investment cost in the electric utility equipment might be expected to have closely followed the Handy-Whitman Index.

Edison Power Company's construction expenditures for production facilities in 1958 averaged $135 per kilowatt.   As determined by the Handy-Whitman Index, such expenditures in 1968 would have averaged about $150 per kilowatt if the company had not taken advantage of larger generating units and other opportunities for economy.   The actual 1968 figure is about $131 per kilowatt.   The difference between the two, $19 per kilowatt, represents largely the effect of the economies described above.

Operating Expenses

Some of the main expenses in generating electricity will be examined to see what the trends are. In each case, the total expense will be shown as cost per unit. For example, the production cost varies almost directly with the units of electricity generated; it is therefore shown as mills per kilowatt-hour. General expense varies with gross operating revenues; it is shown as cents per dollar of revenue. Customer accounts expense usually varies in proportion to the number of customers served; it is shown as dollars per customer. Distribution expense varies to some extent with the number of customers served and kilowatt-hours sold.

Production Cost. Generally speaking, steam plant efficiency is higher when the steam turbines run at high temperatures and pressures. The newer equipment is designed to take advantage of this. Chart 5.14 shows the maximum steam pressures and temperatures used by the electric power industry for 1916, 1930, and 1959.

The aim of the engineer is to generate electricity with the smallest possible number of heat units, in order to cut down the cost of fuel. (The relative efficiency of steam generators can be measured in terms of the pounds of coal it takes to make a kilowatt-hour. The same relative efficiencies would apply in the case of gas or oil or lignite. By using a mathematical formula, the energy in these fuels can be expressed as an equivalent number of pounds of coal.)

Chart 5.15 shows the pounds of coal required to make a kilowatt-hour from 1892 through 1968. In 1892 it took about 8 pounds of coal to make a kilowatt-hour. In 1968, on the average, it took 0.87 pound of coal to make a kilowatt-hour. The new, large, high-pressure and high-temperature units can now make a kilowatt-hour with less than two-thirds of a pound of coal.

Although it takes less coal to make a kilowatt-hour today,

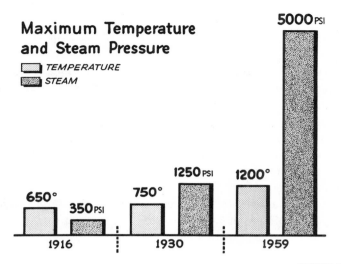

CHART 5.14

the price of coal has been rising over the long term. The two factors have about canceled out. The rise in fuel cost from 1940 to 1948 was caused by the use of old and inefficient machines to meet the demand when generating equipment was

CHART 5.15

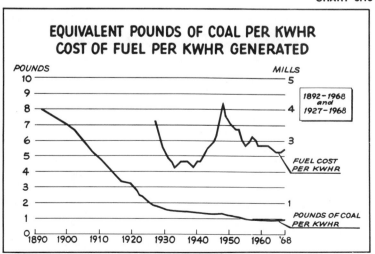

in short supply.   In 1947, the new postwar generators came into operation, and the average fuel cost per kilowatt-hour began to drop gradually after 1948.

Chart 5.16 shows the trend in total production expense of Edison Power Company as compared with the production cost in mills per kilowatt-hour sold.

Administrative and General Expense. This expense is illustrated in Chart 5.17. The total spent by Edison Power Company for administrative and general expense has almost doubled over the past ten years, owing to both inflation and increased business.

Since the amount of these expenses, by and large, varies with sales volume, these expenses are also shown on the chart as cents per dollar of revenue. In these terms, this expense has been dropping slightly. This is the result of two offsetting economic factors. Costs have risen because of inflation, but the increased volume of business has made it possible to use more efficient methods. Personnel has not increased in proportion to increased sales. Without the inflation, this expense would show a greater decline with the increased volume of business, when expressed on the basis of percent of gross revenue.

CHART 5.16

PRODUCTION EXPENSE

Edison Power Company

TRENDS

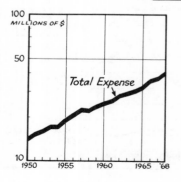
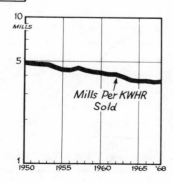

ADMINISTRATIVE AND GENERAL EXPENSE
Edison Power Company
TRENDS

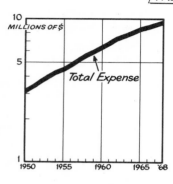

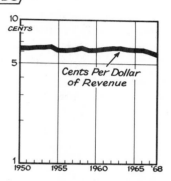

CHART 5.17

Customer Accounts Expense.   Chart 5.18 shows that customer accounts expense has been rising over the past ten years. Edison Power Company has been striving to hold this expense down by modernizing its accounting practices.   It has installed electronic billing machines and automation.   These have helped some, but not enough to hold down the upward trend in the expense as expressed in dollars per customer.

CHART 5.18

CUSTOMER ACCOUNTS EXPENSE
Edison Power Company
TRENDS

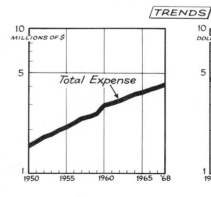

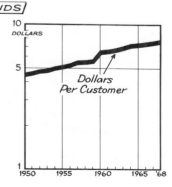

Without the machines the increase in cost would have been greater.

Distribution Expense. As shown in Chart 5.19, here again the total expense has been rising, as well as the expense per customer. This is another case where there have been some economies which have helped offset the effect of inflation, but they have not been sufficient to hold the expense level on a per customer basis.

Transmission Expense. This is shown in Chart 5.20. Sometimes this transmission expense is expressed on the basis of the investment in transmission lines. On this basis the expenses have been dropping, with the exception of one year, since 1952. In this case, going to higher voltages has brought about economies large enough to offset the inflationary effects.

Sales Expense. This expense both as a total and as cents per dollar of gross revenue is shown in Chart 5.21. Before World War II, in 1940, Edison Power Company was spending about 2.2 cents per dollar of its gross revenue for sales expense. This dropped during the war years for the reason that

CHART 5.19

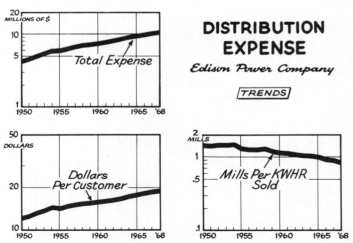

DISTRIBUTION EXPENSE

Edison Power Company

TRENDS

TRANSMISSION EXPENSE

Edison Power Company

TRENDS

CHART 5.20

the company could not purchase generating equipment and the power supply was tight. Under these conditions there was no point in spending money to get more business. After the war, when generators again became available, the company resumed its sales activity and increased its expense to

CHART 5.21

SALES EXPENSE

Edison Power Company

TRENDS

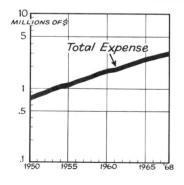

about 1.6 cents for each dollar of revenue in 1946. The company has gradually increased its sales expense to about 1.8 cents for each dollar of revenue in 1968.

Taxes. Taxes have become the greatest single item of expense to Edison Power Company. Sometimes taxes are expressed as a percent of gross revenue; at other times they are expressed as a percent of plant account.

Chart 5.22 shows the taxes of Edison Power both in total dollars and as a percent of sales for the past nineteen years. These taxes have been about 21 percent of gross revenue over the past four years. For the year 1968 they are as follows:

Federal income taxes: 11.4 percent of total gross revenue

Other taxes: 9.8 percent of total gross revenue

Taxes are an expense; like all other expenses, they enter into the cost of furnishing service. In any business, the price to the customer must cover all of the costs or the business will fail. Thus it is the customer who pays these taxes. The company in effect merely acts as tax collector for the government.

Notice in Chart 5.23 that total taxes in 1968 were about 4.5

CHART 5.22

TOTAL TAXES

Edison Power Company

TRENDS

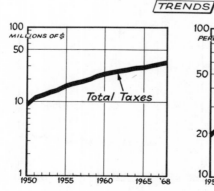
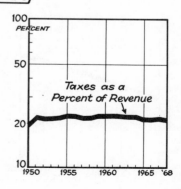

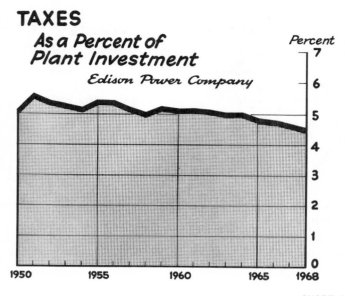

TAXES
As a Percent of
Plant Investment
Edison Power Company

Percent

CHART 5.23

percent of the investment in the Edison Power Company. These are broken down as follows:

Federal income taxes: 2.42 percent of plant investment

Other taxes: 2.07 percent of plant investment

Edison Power Company's taxes are almost equal to the total return on the investment. In other words, to earn the return required to keep the business concern in operation, the company must earn twice that amount in order to pay total taxes and return. The company collected almost as much for the government as it earned for its owners.

However, in comparing tax expense, it is not always correct to compare taxes as a percent of gross revenue. To show the possibility of error, consider company A in Chart 5.24, which pays 21 cents of its gross revenue in taxes, and company B, which pays no taxes. Assuming that all other operating conditions are the same, this is the situation illustrated in the chart.

If company A needs $1 to meet all of its expenses including

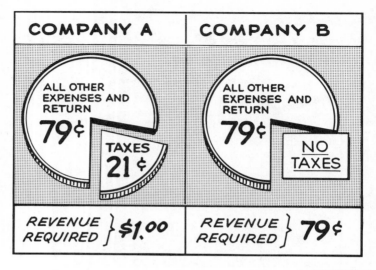

CHART 5.24

taxes, company B requires only 79 cents, since it does not have to pay taxes.

Company A's taxes are 21 percent of A's gross revenue. If it is desired to have B pay equivalent taxes, the application of 21 percent to B's revenue will give a wrong answer (21 percent of 79 cents is only 16.6 cents).

The correct way to make B's taxes equivalent is shown in Table 5.2.

TABLE 5.2

A's gross revenue.	$1.00
Less A's taxes.	0.21
A's revenue less taxes	$0.79
Percentage $\dfrac{\$0.21}{0.79}$	27%
B's revenue without taxes	0.79
B's taxes (27% of $0.79)	$0.21

A still better way of making this comparison is to show taxes as a percent of gross plant investment.

Depreciation. Chart 5.25 shows annual depreciation for Edison Power Company in total dollars and as a percentage of the investment. Notice that it is nearly 2½ percent.

Cost of Money. Because of the high investment and because of the rate of expansion of the industry, the company does not have enough money left over from its annual income to build new facilities out of earnings. For this reason, the company must constantly obtain additional capital in the market. In order to obtain these funds, companies must pay the market price of money. The price the company has to pay to get people to lend it money has to be in line with what investors can get for the use of their money in other enterprises of similar risk.

Chart 5.26 shows the average yield of utility bonds and preferred and common stocks from 1928 through 1968.

Naturally, the higher cost of bond money as experienced in recent years is having its effect on the over-all cost of new money. This, in turn, will have its effect on the over-all re-

CHART 5.25

ANNUAL DEPRECIATION

Edison Power Company

TRENDS

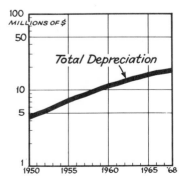
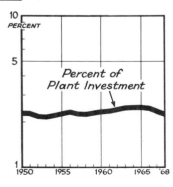

AVERAGE YIELD OF UTILITY BONDS
PREFERRED AND COMMON STOCK

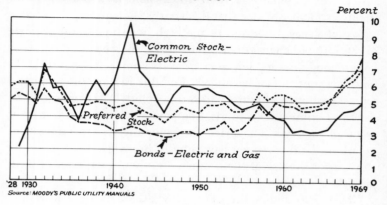

Percent

CHART 5.26

turn that Edison Power Company must earn in order to attract the new capital required for expansion. These are factors the regulatory bodies take into account in determining fair return for rate-making purposes.

6

Forecasting

Forecasting—looking ahead—is essential for all business. It is especially important in the electric utility business. As it takes at least four years to plan and install generating equipment under the most favorable conditions the power company must have a good idea of its customers' needs at least that far ahead.

There are many other reasons for forecasting. Financing is usually scheduled well in advance on a long-term basis. System planning is carried out on a five- to ten-year basis. Rates are affected by changes that may take place in the company's service area. Sales programs, devised to improve the company's position, are based on forecasts.

A forecast to 1980 for Edison Power Company will illustrate some of the factors that are taken into account in looking ahead to the future.

First the company will want to know about the general economic climate. In predicting the general economic climate, it is necessary to make a few basic assumptions. For example, in this case it is assumed:

1. No major war will occur during the forecast period.

2. There will be no extreme fluctuations in economic activity, although there will be minor booms and recessions.

3. The government will continue to spend a sizable amount of money on defense.

4. The public will continue to favor the American free-enterprise system.

5. The regulatory processes will continue, as in the past, to work for the proper interest of all parties.

General economic trends can be determined from figures published by the government, using such indicators as population, labor force, gross national product, households, personal disposable income, construction, and industrial production. These basic data are then used to project or forecast the future economic climate.

Knowing the general economic climate, the company's economists can evaluate local conditions to round out the picture.

Factors in Trending

Among other things, trending involves a study of past patterns. These patterns tend to repeat themselves in the future. Sometimes the past pattern shows seasonal or cyclical fluctuations. Sometimes the past pattern is a curve, sometimes a straight line.

Electric power industry growth has been following a pattern, especially since the late 1920s. Until then electric service was being extended to many customers who had not received it before. This resulted in rapid growth of electric energy sales. By about 1930, practically all communities and most homes in the community had electric service, although many of the farms were not then electrified.

The past pattern of growth in the electric utility business indicates that the industry is growing at a compound rate. The curve of a compound rate of increase is called an *exponential curve*. In such a curve, each figure exceeds the one before it by a certain percentage. Thus a 3 percent compound rate applied to 100 would give 103 for the succeeding year and 106.09 for the next year (103, plus 103 × 3 percent).

Semilog Graph

A compound growth rate is the type of curve likely to be found in the electric power industry. Because it is constantly curving upward, this kind of curve is rather hard to work with on ordinary graph paper. It is hard to predict accurately where the curve is likely to go in the future. It is much easier to make a forecast when the pattern is a straight line. Chart 6.1 shows the same data presented on an arithmetic graph as compared with a semilogarithmic graph.

Using semilog paper, it is possible to convert the curved

CHART 6.1

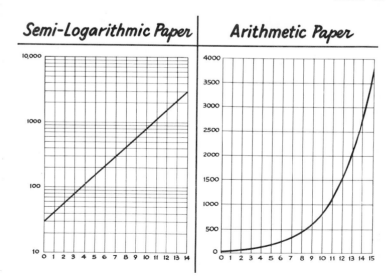

line into a straight line. This paper has a graduated scale, with spaces becoming smaller as the readings move upward. (A semilogarithmic scale graph has a logarithmic scale for the vertical axis and an arithmetic scale for the horizontal axis.) Thus the scale of the paper has the compound rate of increase built into it. In other words, the curve has been taken out of the trend line and put into the scale of the paper. In this manner the economic analyst can deal with straight lines, which makes his task much easier. Semilogarithmic paper will be used in the forecasts in this chapter.

Population

In order to forecast the production and the consumption of goods, it is necessary to have an idea of the size of the future population and the labor force. These data have gone up at a steady rate in the past, and can be projected to future years with considerable accuracy. For example, the labor force for 1990 is already born. Birth and death rate tables, like those used by insurance companies, can be used to predict the number of people between the ages of eighteen and sixty-five twenty years from now. Chart 6.2 shows the estimate of population and labor force by 1980.

In the future, women may make up a larger portion of the labor force. There is a trend toward retirement at an earlier age. Also, more people are going to college and do not start work until later in life. These are only a few examples of the variables which must be taken into account in forecasting the size of the labor force.

Gross National Product (GNP)

The gross national product measures the nation's output of goods and services in terms of its market value. The gross national product is based on three factors at work in the economy. (There is a fourth factor, net export of goods and

POPULATION AND LABOR FORCE

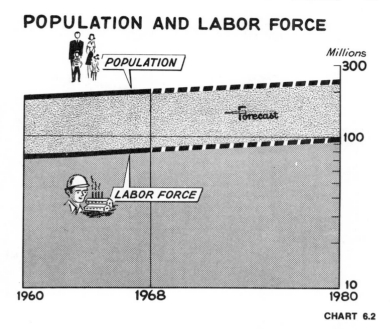

CHART 6.2

services. For present purposes, its effect is negligible.) Personal consumption expenditures represent about 62 percent of the GNP, while gross private domestic investment accounts for 15 percent. The balance, about 23 percent, reflects government spending.

Chart 6.3 shows the gross national product from 1960 through 1968 with a forecast to 1980. It is desirable to express gross national product in constant dollars so that it can be compared from year to year. This eliminates the effects of inflation. A certain year is taken as the base year, and the dollars of all other years are expressed in terms of the value of the dollar in the base year.

Households

The number of households is an important factor in forecasting the probable future demand for electricity, especially

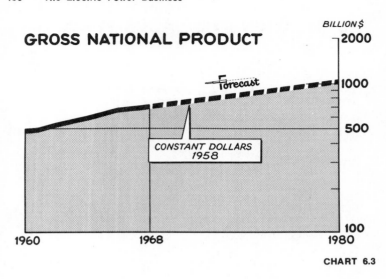

GROSS NATIONAL PRODUCT

BILLION $

Forecast

CONSTANT DOLLARS 1958

1960   1968   1980

CHART 6.3

CHART 6.4

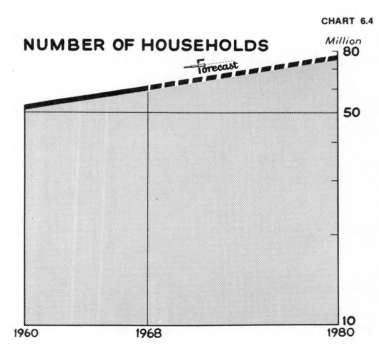

NUMBER OF HOUSEHOLDS

Million

Forecast

1960   1968   1980

in the residential field.   Chart 6.4 shows the forecast of total households.

Personal Disposable Income

Personal disposable income is the amount of money people have to spend.   In 1967 Americans spent about $7.5 billion (or 1.4 percent) of their total personal disposable income for household electricity.

Americans spent $7.1 billion during the same year for kitchen and other household appliances.   This was 1.3 percent of their disposable income.   Relatively speaking, the electric utility industry receives only a small portion of the consumer's income.   Chart 6.5 shows a forecast of personal disposable income in 1958 dollars.

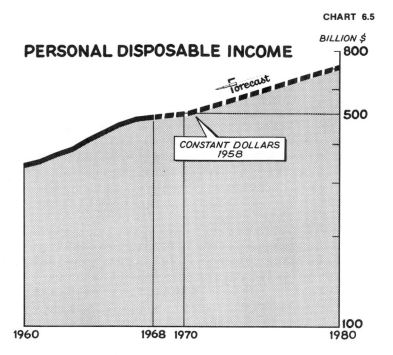

CHART 6.5

BILLION $

PERSONAL DISPOSABLE INCOME

800

Forecast

500

CONSTANT DOLLARS
1958

100

1960   1968 1970   1980

Industrial Production

Another sensitive indicator of national economic activity is the industrial production. It measures the monthly changes in output for the manufacturing, mining, and utility industries. Knowing the trend of industrial production can be helpful in forecasting sales of electricity to industrial customers. Chart 6.6 shows this projection based on past trends.

Patterns of Energy Use

The student of forecasting makes a study of past statistics and performance in all areas to determine whether there are specific patterns. Sometimes the patterns appear so clearly that a mathematical formula can be derived to express them. These formulas are helpful to Edison Power Company in making its forecast to 1980.

The Gompertz Scales. The statistician Gompertz [1] developed a mathematical formula for expressing a kind of long-term growth trend which has been especially helpful in forecasting population. It has also been helpful in the long-range trending and forecasting of commodities and services.

When plotted, the Gompertz curve takes the shape of an elongated "S." Growth begins slowly, then gradually rises more rapidly, after which it gradually levels off, finally reaching a saturation point.

[1] "Gompertz, Benjamin, English mathematician and actuary: b. London, England, March 5, 1779; d. there, July 14, 1865. In 1806 he began to publish a series of papers on advanced mathematical problems which eventually (1819) won him membership in the Royal Society. After the Astronomical Society was founded (1820), he was elected to its council, and in 1822 began to work with Francis Baily on a series of calculations which provided the foundation for the society's catalogue of stars. As actuary to the Alliance Assurance Company he developed a theory of the rate of mortality known as Gompertz' law, which proceeds on the assumption that resistance to death automatically decreases as age increases." *The Encyclopedia Americana*, p. 43, 1968.

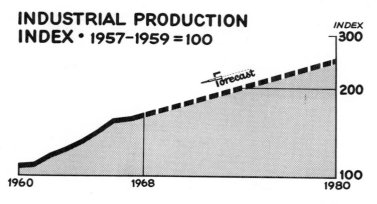

INDUSTRIAL PRODUCTION
INDEX • 1957-1959 = 100

CHART 6.6

There is an actual record of almost ninety years of growth for the electric industry. The industry's forecast to 1980 is likely to have a high degree of accuracy. Taken together, then, there are some 100 years of reliable statistics to which the Gompertz formula can be applied.

Chart 6.7 shows the Gompertz curve fitted to kilowatt-hours per capita in the United States.

Chart 6.8 is a blowup of the portion of Chart 6.7 from 1930 to the year 2000.

A study based on these curves indicates that there is a gradually declining rate of growth in the pattern, as shown in

CHART 6.7

GOMPERTZ CURVE FITTED TO KWHRS PER CAPITA
UNITED STATES

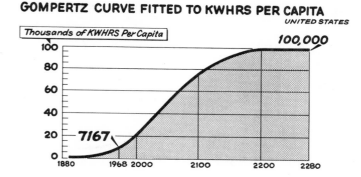

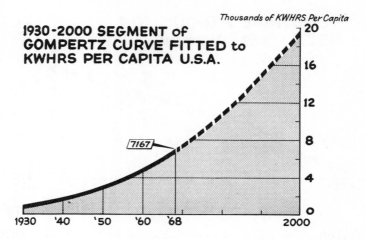

CHART 6.8

Chart 6.9. It shows a high rate of growth during the early years and reaches saturation at about 2200.

An engineer named Fremont Felix has published a statistical study along somewhat the same lines. He examined the personal income, total energy use, and electric use of some

CHART 6.9

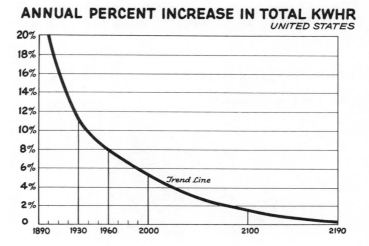

ENERGY and INCOME / *WORLD 1961*

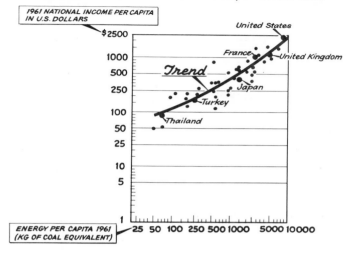

1961 NATIONAL INCOME PER CAPITA IN U.S. DOLLARS

$2500
1000
500
250
100
50
25
10
5
1

United States

France
United Kingdom

Trend

Japan

Turkey

Thailand

ENERGY PER CAPITA 1961 (KG OF COAL EQUIVALENT)

25 50 100 250 500 1000 5000 10000

CHART 6.10

150 nations. As might be expected, he found a correlation between a nation's energy use per capita and its income per capita (Chart 6.10). Inanimate energy use per capita is an index of the amount of machinery being used by the people. The more machinery used to supplement human and animal labor, the higher the production per capita, and the higher the production the higher the income. Felix found the same correlation with respect to electric energy used per capita and personal income. Again, the Felix studies show a decrease in rate of increase in the use of energy. That is to say, the countries having a higher use of energy per capita have a lower rate of increase than those countries using less energy per capita. This is shown in Chart 6.11 for total energy and in Chart 6.12 for electric energy. This same kind of trend is found in electric energy use per capita in the United States (Chart 6.13).

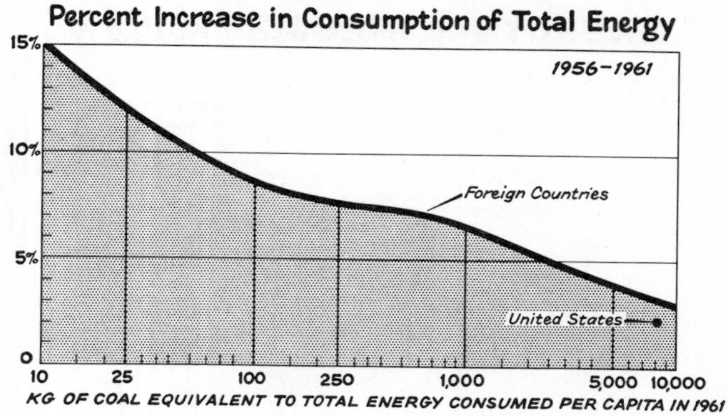

Percent Increase in Consumption of Total Energy
1956–1961
15%
10%
5%
0
Foreign Countries
United States
10
25
100
250
1,000
5,000
10,000
KG OF COAL EQUIVALENT TO TOTAL ENERGY CONSUMED PER CAPITA IN 1961

CHART 6.11

Considering Local Conditions

In addition to the study of the national climate, the Edison Power Company also makes a study of its local conditions. Here are some of the factors the company has considered:

1. Local population trends and trends in labor force
2. Trends in local purchasing power

CHART 6.12

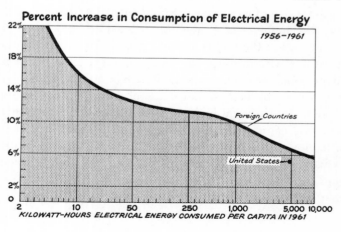

Percent Increase in Consumption of Electrical Energy
1956–1961
22%
18%
14%
10%
6%
2%
0
Foreign Countries
United States
2
10
50
250
1,000
5,000
10,000
KILOWATT-HOURS ELECTRICAL ENERGY CONSUMED PER CAPITA IN 1961

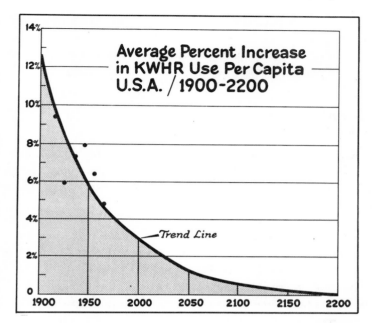

Average Percent Increase in KWHR Use Per Capita U.S.A. / 1900-2200

Trend Line

CHART 6.13

3. The kinds of industries that may be relocating in the territory or leaving the area entirely

4. Conditions in the territory which would attract industry, and what types of industry will be likely to move into the area

5. Future plans of principal industrial customers and large employers concerning expansion or contraction

6. Sales prospects for electric appliances and equipment

7. The future needs of the customers in the several company divisions

8. The trends in kilowatt-hour requirements and kilowatt requirements for all classes of service, by communities, by districts, and for the company as a whole

9. The company's rate policy and its effect on the customer's use of electricity

10. The company's sales program and the company's future selling plans

11. The company's area development and industrial development plans

Forecasting Electricity in the United States

Every power company periodically makes the kinds of forecasts described here as part of its over-all planning. From time to time these forecasts are reviewed and changed to meet changing conditions.

In 1959 an exhaustive forecast was made of the total electric utility industry. This forecast was made two ways: by a task force with knowledge of its local conditions and by the staff of the Edison Electric Institute.

In order to supply the information for the forecast by regions, the task force made estimates of the peak loads and energy requirements in its respective regions. The Edison Electric Institute assembled all the regional forecasts and summarized the industry's forecast based on local conditions. The staff of the Institute then made a forecast based on a number of correlations. Kilowatt-hour sales on a national basis are influenced by the level of business activity, the growth of the economy, and population. In analyzing historical trends in kilowatt-hour sales it was found that positive relationships existed between kilowatt-hour sales in the various customer classifications and components of the gross national product in constant dollars, and the Federal Reserve Board Index of Industrial Production. By a series of correlations it was found that long-term estimates of kilowatt-hour sales could be derived with a reasonable degree of confidence from estimates of growth in the national economy as measured by GNP in constant dollars. Residential sales showed a definite correlation with disposable personal income; commercial sales correlated with personal consumption expendi-

tures for services; and industrial sales correlated with the
FRB Index of Industrial Production (Charts 6.14, 6.15, 6.16).
All these, in turn, could be correlated with GNP in constant
dollars. Two such estimates were made. One assumed the
GNP (in constant dollars) would continue to grow at the 3.57
percent average annual rate of the twelve years 1946–1958,
and the other assumed an annual average increase in the
GNP of 3 percent per year, which was the average rate of
growth in GNP (in constant 1954 dollars) over the preceding
fifty years.

The aggregate kilowatt-hour generation as estimated by the
task force coincided with the projection of kilowatt-hours

CHART 6.14

Correlation Between Disposable Personal Income
(1954 Dollars) and Residential KWHR Sales

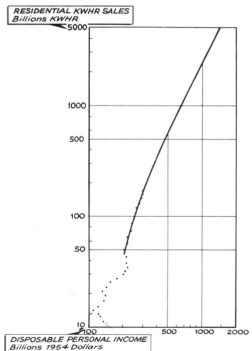

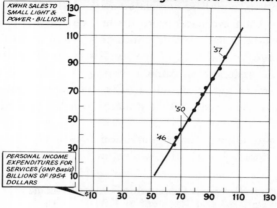

CHART 6.15

based on the average rate of growth in GNP between 1946–1958 of 3.57 percent per year compounded. Generation was estimated from total sales by adding an estimated amount for losses in transmission, company use, and so on. The projection was based on a 3.57 percent rate of growth in GNP. To project to the year 2000 involved a great deal of speculation. Studies by EEI indicate that in the year 2000 total output will range between 8 trillion and 10 trillion kilowatt-hours, depending on the rate of growth of the national economy and the rate at which new energy uses are introduced, especially electric space heating (Chart 6.17).

Every year the forecast is checked against the actual kilowatt-hour production for that year. The cumulative error has been so slight that it has been found unnecessary to change the forecast from that made in 1959. Through 1968 the cumulative error was 0.182 percent—slightly less than two-tenths of one percent. That is, the actual production through 1968 was within about two-tenths of one percent of the production forecast nine years ago. On this basis the cumulative error for 1980 should be within the range of 1 percent, plus or minus.

Correlation Between Industrial KWHR Usage and FRB Industrial Production Index

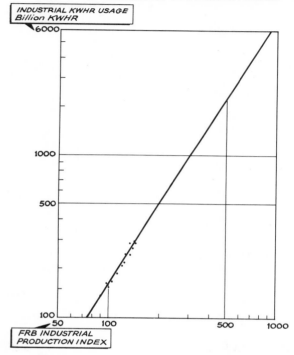

CHART 6.16

Trends of the Edison Power Company

Kilowatt-hour Sales. Chart 6.18 shows a forecast of Edison Power's sales based upon its past record. In 1980 the company expects sales of about 23.3 billion kilowatt-hours as compared with 10.8 billion kilowatt-hours in 1968. This is a compound growth rate of about 6.6 percent a year. The company should almost double its total sales in about ten years.

Other Patterns. Once having determined the basic forecast of kilowatt-hours, the analyst then proceeds to forecast kilowatts of demand, plant investment, gross revenues, various

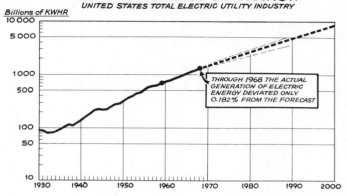

ELECTRIC ENERGY GENERATION
UNITED STATES TOTAL ELECTRIC UTILITY INDUSTRY

THROUGH 1968 THE ACTUAL GENERATION OF ELECTRIC ENERGY DEVIATED ONLY 0.182% FROM THE FORECAST

CHART 6.17

expenses, and the like. These forecasts are influenced by the price of service and that price, in turn, is influenced by the efficiency of operation. They show the efficiency of operation, that is, the efficiency of converting raw fuel to electric

CHART 6.18

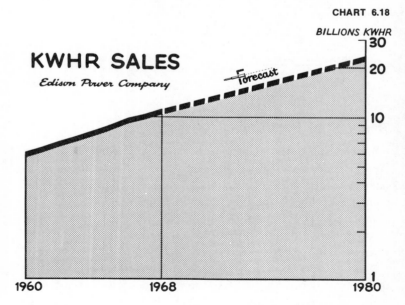

BILLIONS KWHR

KWHR SALES

Edison Power Company

Forecast

1960   1968   1980

energy. Again, the over-all patterns developed in the past can be used in more accurately predicting the future.

In general, the economy of scale usually influences most enterprises. It is especially influential in the conversion and delivery of energy. Generally speaking, the larger the generating unit, the higher its efficiency. The higher the transmission line voltage the more energy per dollar of investment it can carry. Consequently, research and development effort are directed toward developing ways of bringing about higher efficiencies.

Chart 6.19 forecasts the pattern of growth in the size of generating units. Chart 6.20 shows the pattern in maximum transmission voltages. Chart 6.21 shows how the investment per kilowatt in generators goes down with size. (Chapter 11 includes a discussion of coordination and pooling and shows how companies are utilizing the economies of scale.)

Forecast of Revenue and Expenses

Gross Revenue. Gross revenue is affected by the declining steps in the electric rates. In forecasting gross revenue, merely extending the curve of past sales will not give the

CHART 6.19

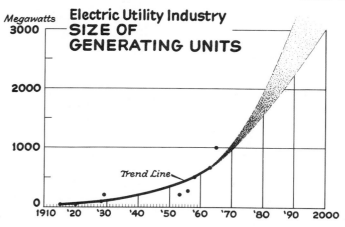

Megawatts **Electric Utility Industry**
3000 **SIZE OF GENERATING UNITS**

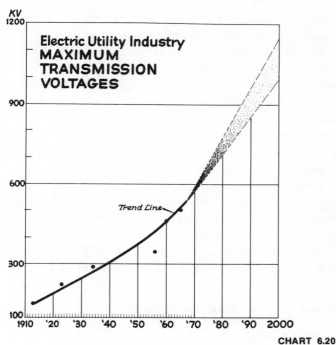

CHART 6.20

CHART 6.21

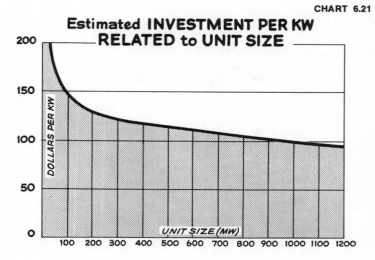

right answer.   There are other elements which must be considered if a reasonably precise forecast is to be made.   Here is
one way such a forecast can be made:

To forecast gross revenue, it is necessary to know how
many kilowatt-hours will be sold and the price per kilowatt-
hour.   A forecast of kilowatt-hour sales can be made for each
class of service in the same manner as the forecast shown in
Chart 6.22.   Making a forecast of the average rate for each
class gives the price per kilowatt-hour.   Because of increased
use of electricity, more is being sold at the lower steps in the
rate schedule.   This has produced a trend which can be forecast with reasonable accuracy.   Of course, if rate changes are
expected during the forecast period, the new rates will have
to be taken into account.

This average rate per kilowatt-hour for the class is then applied to the forecast kilowatt-hour sales for each class.   This
gives the projected revenue for each class.   The total revenue
for the company can be found by adding up the revenues of
the various classes.

CHART 6.22

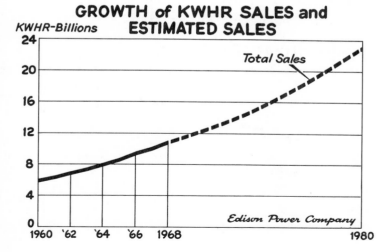

GROWTH of KWHR SALES and ESTIMATED SALES

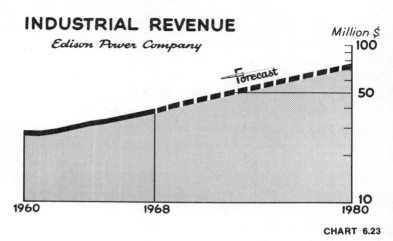

INDUSTRIAL REVENUE
Edison Power Company

Million $

100

Forecast

50

1960   1968   1980

10

CHART 6.23

These class revenues are shown in Charts 6.23 to 6.25. Chart 6.26 shows the forecast of total gross revenue—the sum of the revenue of the classes, including any other additional revenues.   (Chapter 10 discusses another method of determining revenue from kilowatt-hours without going through the above process.)

Investment.   Chapter 5 outlined the trends in investment per kilowatt for the production, transmission, and distribution

CHART 6.24

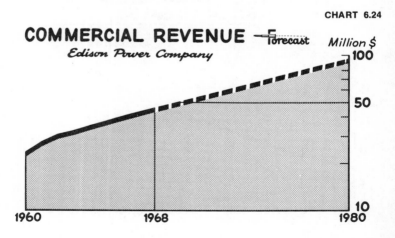

COMMERCIAL REVENUE —Forecast   Million $
Edison Power Company

100

50

1960   1968   1980

10

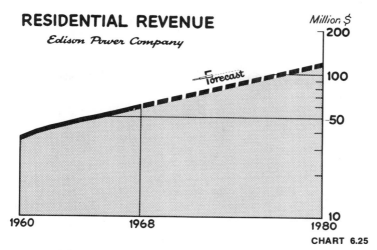

RESIDENTIAL REVENUE

Edison Power Company

Million $

CHART 6.25

systems of the company.   Chart 6.27 shows how total investment per kilowatt has been running, with a projection into
the future.   From this projection it is possible to find the total
investment projected for Edison Power Company in 1980.

CHART 6.26

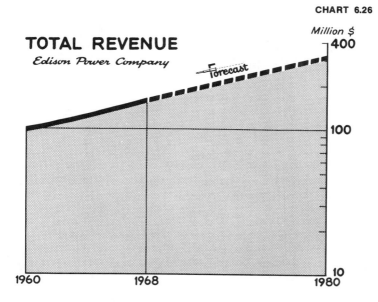

TOTAL REVENUE

Edison Power Company

Million $

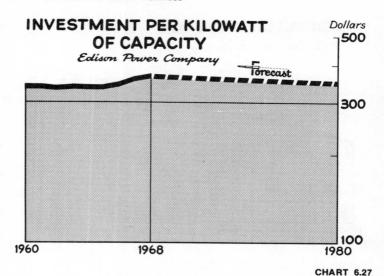

CHART 6.27

Operating Expenses. The operating expenses include all costs incurred in furnishing electric service and exclude depreciation, taxes, and return on investment. The trends in these expenses are shown below to indicate the bearing they will have on the future operation of the company.

Production Expense. Edison Power Company shows a continuing rise in production expense principally because of greater generation. Increases in fuel prices and labor rates have been more than offset by improved efficiencies (Chart 6.28).

Transmission Expense. While total transmission expense has been rising, this cost in cents per dollar of transmission investment has been going down. This shows the offsetting economies which have come from the use of higher transmission voltages. The trend of this expense is shown in Chart 6.29. It is expected to continue at the same pace in the future.

Distribution Expense. Distribution expense is rising, both in total dollars and in dollars per customer. On a kilowatt-hour basis the cost trend is down. There have been fewer

PRODUCTION EXPENSE

Edison Power Company

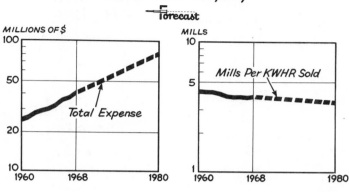

CHART 6.28

opportunities for cost-cutting measures in this field than in some of the others.

The trend of distribution expense is shown in Chart 6.30. The economists predict that this cost will continue to rise on a dollar per customer basis.

Customer Accounts. After a careful review of the company's plans to install more electronic bookkeeping machines,

CHART 6.29

TRANSMISSION EXPENSE

Edison Power Company

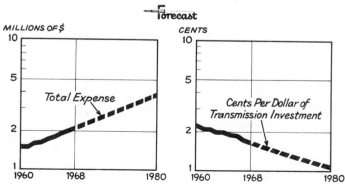

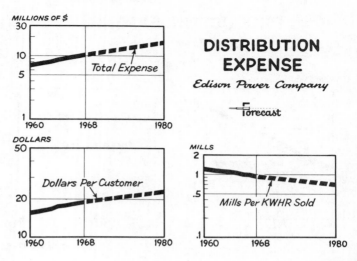

CHART 6.30

the company economists believe that the trend in customer accounts expense will be shown in Chart 6.31.

Sales Expense. The total dollars spent in sales promotion will increase over the next ten years, and it is estimated there will be no increase in cents per dollar of gross revenue. The

CHART 6.31

CUSTOMER ACCOUNTS EXPENSE

Edison Power Company

Forecast

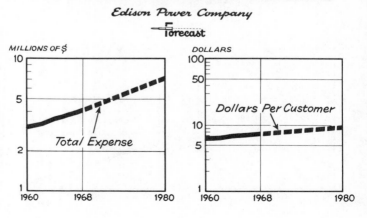

amount of this expense item hinges largely on the kind and scope of the company's sales program. This is discussed in greater detail in Chapter 7. For a view of the trend and forecast of this expense item, see Chart 6.32.

Administrative and General. The company's economists feel that this expense will be slightly lower over the predictable future, when expressed in terms of cents per dollar of revenue. This projection is shown in Chart 6.33.

All of these various expense trends are shown in Chart 6.34.

Summary. In this way the company makes forecasts of all the factors that management needs to know for planning purposes. The forecasts serve as a guide to management in determining future policy. They help management determine when to buy new equipment and in what sizes. They help decide where new plants will be located. They help the company schedule its financing program, and estimate future earnings and percent return. They guide management in estimating the number and kind of employees the company will need for the future—how many engineers, linemen, salespeople, operators. They may point out ways to cut expenses and new methods of doing things. The forecast helps guide the

CHART 6.32

SALES EXPENSE

Edison Power Company

Forecast

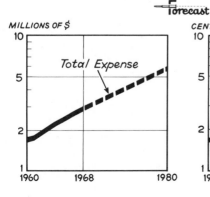
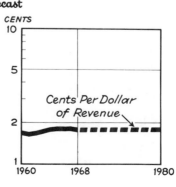

ADMINISTRATIVE and GENERAL EXPENSE

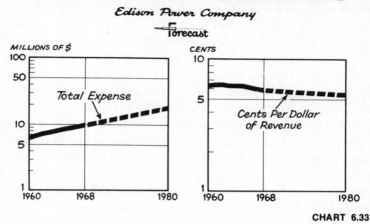

CHART 6.33

sales program.   Management can figure out how much the company can afford to spend to get new business.   The forecast will help to determine how the sales program can be corrected if sales are lagging in a certain field.

Finally, the forecast will be a guide in working out the

CHART 6.34

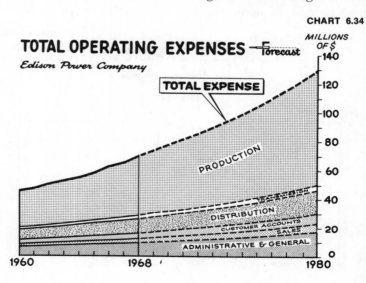

company's rate policy.   It will help show when and how much the rates need to be changed.

The forecast is an estimate.   It is no better than the knowledge and judgment of those making it.   No one can predict the future with certainty, but experience in this field enables a company to make a reliable forecast.   It is an excellent tool for management.

Forecasting is a much more complicated and comprehensive procedure than indicated here.   A book could be written on this subject alone.   The purpose here is merely to touch on some of the major principles.

7

Getting New
Business—Selling

Electricity is a convenient and flexible source of energy for home, farm, business, and industry. Nevertheless it has to be sold. People must be informed of the many uses of electricity. The Edison Power Company uses many means to acquaint its customers with the benefits of electricity. Where electricity can do a job better, the way is pointed out to the customer. This may be done in many ways—by advertising, through a personal call by a sales representative of the power company, or by a dealer in electric appliances.

Increased sales enable the company to take greater advantage of the economies of mass production. This results in a lower unit cost and price of electricity. The lower price in turn tends to increase sales.

Sales of electricity have been doubling about every ten years. Chart 7.1 shows this increase for residential use from 1915 through 1968.

Chart 7.2 shows how much the American people spend for electricity in the home, as compared with other items. The use of more and more electricity in the home and in industry means a rising standard of living. Although the population is growing, the use of electricity is growing more rapidly. The total kilowatt-hours sold in 1968 was over twice the sales in 1958, while population increased only 15 percent during the same period. The number of kilowatt-hours per capita used in the United States increased from 4,203 kilowatt-hours in 1958 to 7,167 in 1968, or about 71 percent.

Electricity has played an important part in the industrial development of America. In America today, most of the work is done by machinery. Most of this machinery is run by electric power. The power can be delivered directly to the machine and the operator can control it by push button. Because of this machine production in America, every working man has working for him the equivalent of 512 assistants. (It has been estimated that in a working year the average worker uses muscular energy equivalent to 67 kilowatt-hours

CHART 7.1

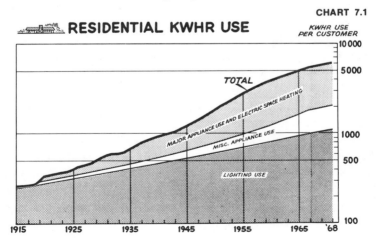

RESIDENTIAL KWHR USE

KWHR USE PER CUSTOMER

TOTAL

MAJOR APPLIANCE USE AND ELECTRIC SPACE HEATING

MISC. APPLIANCE USE

LIGHTING USE

10 000
5000
1000
500
100

1915 1925 1935 1945 1955 1965 '68

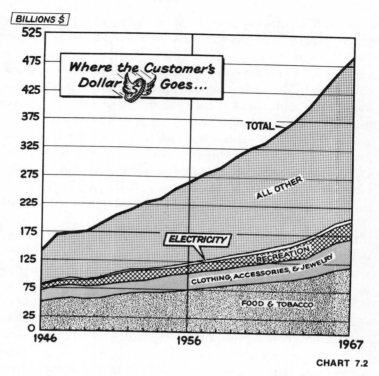

BILLINGS $

Where the Customer's Dollar Goes...

TOTAL

ALL OTHER

ELECTRICITY

RECREATION

CLOTHING, ACCESSORIES, & JEWELRY

FOOD & TOBACCO

1946   1956   1967

CHART 7.2

of electrical energy. In 1967 the annual kilowatt-hours of electricity used by the U.S. worker averaged an estimated 34,336. Dividing 67 into 34,336 gives 512, the number of electrical helpers each worker has.) This is one of the reasons why the productive capacity in America is so much higher than in the rest of the world. Where some other countries rely on manual labor and muscle power, American working-men are the operators of machines.

The growth of America's electric power industry is due in part to popular demand for more and more of the conveniences of electricity. Part of the growth can be attributed to the American free-enterprise system. Part is also due to the selling efforts of power companies, electrical manufacturers, and other electric agencies.

The Competitive Markets

Before discussing how the Edison Power Company conducts its sales program, it will be helpful to have a general discussion of the whole field of energy markets. Remember, the electric utility business is one that converts and upgrades raw energy into electric energy. Traditionally, there has been keen competition among the suppliers of raw energy, such as coal, gas, and oil. A customer may use any one of these to heat his home, business, or factory. Industry may use these fuels to drive steam engines, oil engines, gas engines, or to generate its own electricity by converting them to electric energy. This competition is healthy and works in the public interest.

By and large, there are four primary competitive markets for the use of energy, namely: (1) lighting, (2) stationary motors in homes, business, and factories, (3) heating, and (4) transportation. For reasons previously mentioned it has been possible, as a result of research and innovation, to improve the efficiency of converting raw fuels to electric energy and to bring about a decrease in the average price of electricity sold to the consumer. Those who provide raw fuels are also constantly carrying on research to improve their efficiency in removing fuel from the ground and delivering it to the consumer. However, the opportunities for improvement in the efficiency of these operations are not as great as the opportunities for improvement in the efficiency of converting the raw fuel to electric energy. For these reasons, and because of the increase in electric utilization equipment, the growth of electric energy in the country has been at about twice the rate of the use of raw energy (Chart 7.3).

By the early part of the century most of the lighting in the nation had changed from oil lamps and gas to electricity. Today the lighting market is almost universally electric. Be-

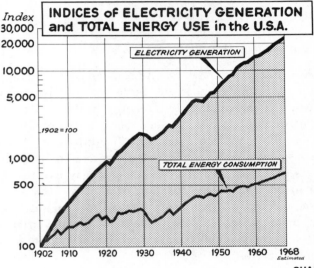

INDICES of ELECTRICITY GENERATION and TOTAL ENERGY USE in the U.S.A.

CHART 7.3

ginning in the 1920s manufacturers and industry began to realize that they could change the gas, oil, and steam engines in their factories to electric motors and obtain electric energy from one of the interconnected power systems, thus obtaining energy more reliably and more economically. As a consequence, there has been a gradual and steady changeover to electric energy in the stationary motor market.

The third major market for energy is heating. Electricity began making inroads in this market through heating appliances such as the toaster, the percolator, the hot plate, the electric range, the water heater, and the dryer. In commercial establishments electric fry kettles, hot plates, and electric ranges appeared. In industry heating applications took the form of large furnaces.

Electric space heating began to be competitive with gas and oil in the last decade as power companies were able to reduce their prices in many areas. Only a few years ago few homes were heated electrically. In 1968 there were almost 3.5 million. In 1980, estimates call for about 19 million. It is estimated that by the year 2000 about 70 million homes in the country will be electrically heated.

In commercial establishments and in industry electric heating has become competitive in many areas. There are numerous all-electric shopping centers, office buildings, schools, churches, and hotels. One innovative technique is to provide the heating by the lighting in the building. A multistory office building, for example, can be heated entirely by the lighting required to give adequate lighting levels in the offices. In the summer the heat from the lighting is removed by ducts and by air conditioning.

Electric energy has been used in transportation for some time. Electric street cars, commuter trains, and railroads are not new. In some countries, such as Switzerland, France, and Japan, railway electrification is more advanced than it is in the United States, but some of the principal railroads in this country are now studying the possibility of converting diesel locomotives to electric locomotives. In the early 1950s a study was made by an independent firm, which concluded that there would be savings to the railroads through electrification. The study indicated such savings would pay about a 10 percent return on the added investment. However, the railroads were not in a position to finance the changeover at the time. Since then the price of electric energy has declined.

There is a large market for energy in automobiles, trucks, buses, and in farm and industrial vehicles. A number of manufacturers are working on battery improvements, and a few electric automobiles are already on the market. Surveys of public opinion indicate that something like ten million people would buy an electric car if it were available at a reasonable cost. Of course, the growing interest in clean air has given the concept of the electric automobile great impetus.

Now that practically everyone in the country has electric energy available, including the farm customers, most of the sales opportunities for electric utility companies come from increasing sales to existing customers. There is also some increased business to be gained from increased population and family formations.

With the gradual trend toward use of electric energy by the consumer, it is understandable that over the years there has been a gradual increase in the amount of raw energy used to generate electric energy. In 1920 about 10 percent of the raw energy consumed was used to produce electric energy, as compared with more than 20 percent today. The figure will be 30 percent by 1980 and somewhere between 40 and 50 percent by the year 2000 (Chart 7.4).

Edison Power Company Sales Program

During the early years of Edison Power Company, electric power was used mainly for lighting. There was a time when power plants ran only during certain scheduled hours in the evening. At that time it was not worthwhile to run generators in the daytime because there was virtually no use for electricity during daylight hours. When other uses for electricity were developed, service was expanded to twenty-four hours a day.

Lighting was the first popular use for electricity in the

CHART 7.4

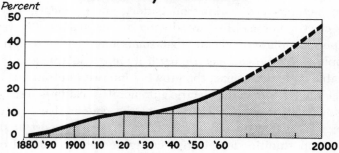

Estimated Conversion of Primary Energy to Electricity 1882–2000

home. The next was the electric iron. Despite what they had read about the electric iron, great numbers of Edison Power Company's customers continued to use the heavy old-style iron from which their mothers had received many a burn. The company's salespeople had to visit customers' homes and prove to them that this newfangled gadget would work, and that it had advantages over the old-style iron.

A few years later, the company again put on demonstrations to show that the electric refrigerator would work, and that it would keep foods fresh. Company men demonstrated the electric sweeper, house to house. The housewife had to be convinced that it would clean rugs. Edison Power salesmen washed clothes for the housewife to show that this could be done by electricity.

The market had to be built and this took salesmanship. These old appliances were quaint, indeed, compared with modern appliances. In turn, it may be expected that present appliances will appear quaint when compared with those of tomorrow.

The first electric ranges were rather crude devices, not nearly as efficient as modern ranges. People had to be convinced that this method of cooking was practical and convenient. To help do this, Edison Power Company gradually built up a home service department which sent ladies to visit customers' homes to help them to get the most out of their electric ranges.

Lighting. Lighting is a good example of the nature of the growing market for electricity. Chart 7.5 shows the lighting use in the home in 1915 and 1968. Even though electric lighting has been on the market since the first days of the industry, there is still great potential for growth.

In the book *Seeing* (Matthew Luckiesh and Frank K. Moss, The Williams and Wilkins Company, Baltimore, 1931) some interesting observations are made on lighting. Primitive man was accustomed to seeing distant objects under the light of thousands of foot-candles produced by the sun. (The foot-can-

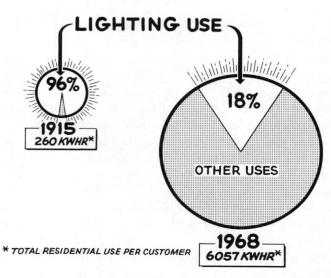

LIGHTING USE

96%

1915
260 KWHR*

18%

OTHER USES

* TOTAL RESIDENTIAL USE PER CUSTOMER

1968
6057 KWHR*

CHART 7.5

dle is a measure of light intensity. It is equal to the intensity of the light of a candle held at a distance of one foot.) Modern man tends to do most of his seeing under a few foot-candles, while doing close-up work such as reading or operating a machine.

Chart 7.6 shows the foot-candles of lighting in the noonday sun on the beach as compared with the foot-candles in the daytime under the shade of a tree. Compare these with the 20 to 150 foot-candles that might be available for reading purposes in the home or office and for work in the factory. The room for improvement is great—and so is the market for electricity in this field. People are beginning to realize that good lighting can help improve efficiency and increase production. Careful experiments show that when there is more light, less energy is wasted on useless work.

Chart 7.7 illustrates that the ease of seeing and the amount of light are directly related. Read the matter in this chart and notice when the reading becomes difficult or impossible. Then hold the book close to the light. One can read more of the material when there is more light on it.

CHART 7.6

CHART 7.7

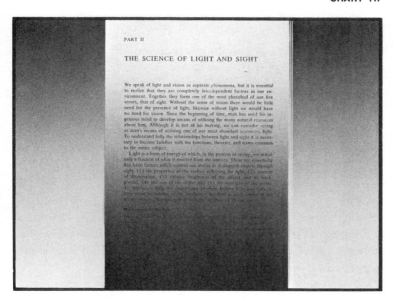

For years, Edison Power Company has been interested in raising the lighting level in the homes, factories, and offices it serves. This is good for both the customer and the company. Every year the company puts on a campaign to sell higher-wattage light bulbs. A couple of the salesmen worked out an ingenious way to show people they need brighter lights. They rigged up a chair with a light which could be made dimmer or brighter by turning a knob. They asked the customer to sit in the chair and read a newspaper. He was asked to turn the knob to the level of light which seemed most comfortable to him for reading. A dial attached to the controls showed the amount of light he had selected. Invariably the person would turn to a lighting intensity five or even ten times as great as he had in his home. After a demonstration like this, the customer usually bought 100- or 150-watt bulbs. On the average, people buy 60- to 75-watt bulbs. In most fixtures, this size does not give the light people want and need.

Edison Power has found that with each campaign to sell higher-wattage bulbs, the whole level of lighting wattage is raised. This in turn raises the level of illumination in the home, office, and plant, to everyone's benefit. The company is also trying to get people to put more outlets in the home so that lighting can be placed where it is most useful and convenient.

The whole field of decorative lighting has almost unlimited possibilities. Edison Power has full-time lighting sales engineers who spend their time working with architects and builders to help them plan adequate lighting levels and to give them suggestions for using lighting as a decorative device.

Quite a few "light-conditioned" homes have been built in Edison Power Company's service area. These are examples of homes provided with all the most modern lighting conveniences. Chart 7.8 shows what a light-conditioned home might use as compared with the average home.

TODAY'S AVERAGE HOME		The LIGHT-CONDITIONED HOME
21.6	Number of lamp bulbs and tubes	**70**
1000	Electrical Consumption KWHR lighting only	**2000-2500**
12	Lighting Fixtures	**20-25** plus wall lighting
7	Number of portable lamps	**15**

CHART 7.8

Refrigeration. Electric refrigeration has reached almost complete saturation in the home. (Saturation is the electric industry term describing the proportion of the total appliance market that has the appliance in use and is usually expressed as a percentage. The change in saturation of an appliance is a rough measure of the effectiveness of the sales program, and also of the remaining potential market for an appliance.) But new improvements in electric refrigeration for the home are constantly being developed. A 12-cubic-foot refrigerator will use about 700 kilowatt-hours a year, but the refrigerator-freezer will use 1,100 kilowatt-hours a year. From a use standpoint, it is like having two refrigerators in the home.

Recently a frostless refrigerator-freezer has been developed. This appliance uses an estimated 1,700 kilowatt-hours per year—2½ times the use of the conventional refrigerator. This is shown in Chart 7.9. All three of these refrigerators have a diversified load factor of about 85 percent, making refrigeration one of the best loads the company has. Refrigerators go on and off at intervals all day and all night. The chances of all of them in the company's service area being on

REFRIGERATION

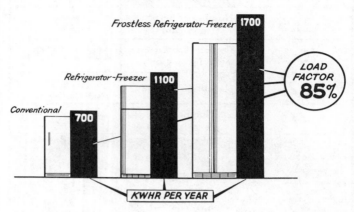

CHART 7.9

at one time are slight. Similarly, they are not all likely to be off at the same time; rather, the demand will tend to average out. As one comes on, another may go off. Thus as a group they will use a relatively steady amount of electricity around the clock, without producing any great demand at a particular hour of the day. Because of this, the company will get high utilization out of the kilowatts of generating capacity used to meet the refrigeration load. (Load factor is described fully in Chapter 2, pages 30 to 34.)

Saturation in Major Appliances

Lighting and refrigeration are only two examples of the expanding market for electricity in the home. Table 7.1 shows the national saturation for some of the major appliances in 1968.

Heating and Cooling. Air conditioning has grown rapidly in recent years. Prior to World War II, air conditioning was used mainly in theaters and stores. The high cost and huge bulk of the machinery required to cool air made it prohibitive for most home owners. After the war, manufacturers

TABLE 7.1　National Saturation for Major Appliances

Appliance	Percent
Air conditioners (room)	42.5
Dishwashers	20.8
Freezers	28.5
Ranges	49.9
Water heaters	27.8
Refrigerators	99.8
Bed coverings	45.6

turned their attention to this problem and began to produce compact air conditioners at moderate prices. Refinements and improvements are still being made today, but the current price, size, and effectiveness of room air conditioners have made them popular in many areas of the country. Chart 7.10 shows sales figures for room air conditioners by years since 1957.

Central air conditioning is also growing. It involves installation of a central cooling plant from which air is circulated to all parts of the home. It can be either a heat pump, which heats in winter and cools in summer, or separate cooling machinery.

CHART 7.10

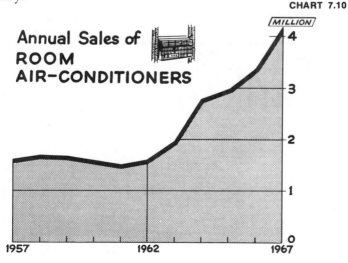

Annual Sales of
ROOM
AIR-CONDITIONERS

This air conditioning load has had a marked effect on load characteristics, especially of companies in the South and Southwest. In most of the companies in those areas, the time of the company's peak load has changed from December to the summer. More important, the summer peak has become much higher than the winter peak had been. This has caused a decrease in the annual load factor of many companies operating in the hot part of the country, as shown in Chart 7.11.

As it is not in the Southwest, Edison Power Company's peak has not yet changed from a winter peak to a summer peak. Since the summer peak has not exceeded the winter peak, Edison Power's annual load factor has not yet suffered from this shift.

Chart 7.12 shows the relationship between temperatures and the company system load for a company operating in the Southwest. Notice that on the cooler days the company's load drops substantially.

In 1968, about 90 percent of all the electric power load in

CHART 7.11

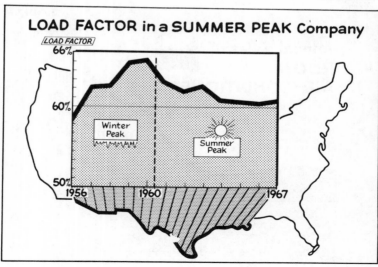

LOAD FACTOR in a SUMMER PEAK Company

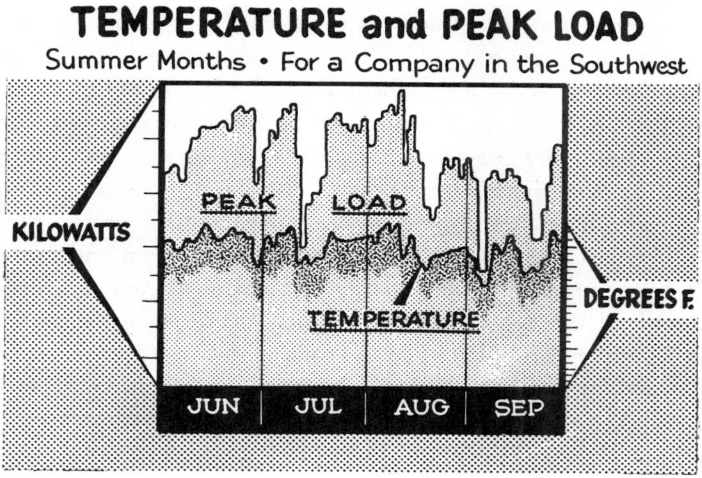

TEMPERATURE and PEAK LOAD
Summer Months • For a Company in the Southwest

KILOWATTS

PEAK LOAD

TEMPERATURE

DEGREES F.

JUN | JUL | AUG | SEP

CHART 7.12

the country had a peak in the summer. Air conditioning is growing more and more popular in the northern part of the country.

Coming also is the control of humidity. For year-round comfort, moisture should be added to the inside air in winter, and moisture should be taken out in summer.

This unusual growth in complete air conditioning is being studied carefully by power companies. They want to be sure to have the capability to serve it. Also this has brought into focus the large market in the field of air conditioning and in the companion field of house heating.

House Heating. Edison Power Company has done considerable work in electric house heating. There are a good many electrically heated homes in the company's service area. Some have radiant panel heating, others the heat pump. Each has its place, based on the particular local conditions.

The heat pump is simply an ordinary refrigeration device which both heats and cools. The refrigerator in the home pumps heat out of the refrigerator into the kitchen. Warm air can be felt coming from the back or bottom of a refrigerator.   This is the heat that was taken out of the inside of the refrigerator to make it cold.   The air conditioner is a refrigeration device which pumps heat from inside the house to the outside.   If the unit is turned around so that the part normally facing inward is made to face outward it will pump heat from the outside into the house.   The heat pump accomplishes both of these functions by pumping heat from within the house to the outside during the summer months and pumping heat from the outside to the inside when heat is needed.   Even when it is quite cold outside there is heat in the air which can be pumped indoors.

As shown in Chart 7.13 this heating and cooling market is about as great as the combined market for all other uses in the home.   The all-electric home uses about 26,000 kilowatt-hours a year.

New Appliances

Chart 7.14 shows that the number of appliances in the market continues to grow.   There were only about nineteen appliances in common use in 1930; now there are around

CHART 7.13

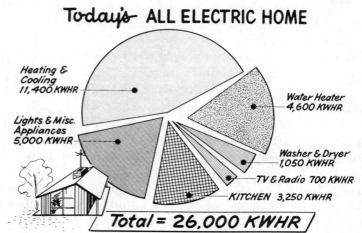

Today's ALL ELECTRIC HOME

Heating &
Cooling
11,400 KWHR

Lights & Misc.
Appliances
5,000 KWHR

Water Heater
4,600 KWHR

Washer & Dryer
1,050 KWHR

TV & Radio 700 KWHR

KITCHEN 3,250 KWHR

Total = 26,000 KWHR

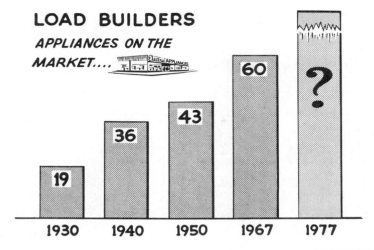

LOAD BUILDERS

APPLIANCES ON THE MARKET....

1930	1940	1950	1967	1977
19	36	43	60	?

CHART 7.14

sixty. New appliances are being developed at the rate of about 1½ per year. Appliances in use today that were not generally in use twenty years ago include television, the washer-dryer, dishwasher, frost-free refrigerator, food freezer, room air conditioner, heat pump, radiant panel heating, and many others, including recent developments in the field of home power tools—portable drills, saws, and the new combination tools. Ten years from now, it is fairly certain that there will be at least fifteen new appliances, some of which are not even in the laboratories today.

The Farm Market

The farm market also offers a large potential for growth. The number of farms has been decreasing, as shown in Chart 7.15, but total electricity sales on the farm have been growing steadily.

In 1940 the average investment per farm was only about $8,000 (Chart 7.16). By 1950, capital requirements had nearly tripled, to about $23,000 per farm. In 1967, the aver-

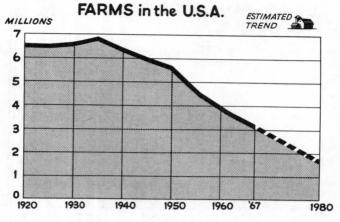

CHART 7.15

age investment was about $86,000. The average size of farms is increasing, and more machinery is being employed. Much of this machinery is in the form of electrically power-driven tools. The progressive farmer is using more and more electrically operated equipment to help him get his work done.

Chart 7.17 shows that farmers are customers for both farm production equipment and home appliances.

CHART 7.16

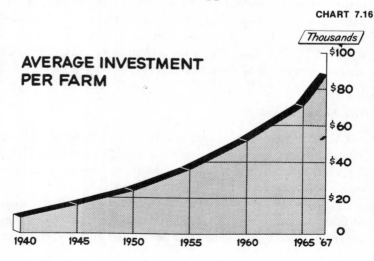

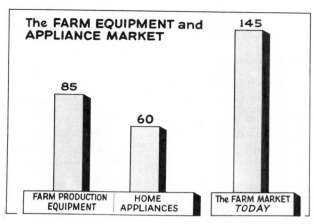

The FARM EQUIPMENT and APPLIANCE MARKET

85 — FARM PRODUCTION EQUIPMENT

60 — HOME APPLIANCES

145 — The FARM MARKET TODAY

CHART 7.17

Mechanization means that the farmer can get more work done and produce more food. By doing so he earns more. What he earns is a combination of administrator's salary, wages for his efforts, and return on his investment. The more he is mechanized the more valuable his time becomes in each of those categories. The more his time is worth, the more he must continue to mechanize, both to conserve his time and to get most use of his existing investment.

Because he has a good deal of capital tied up in his farm, the farmer must get the best possible yield from every farming operation. If environmental control can raise his production, he is a prospect for it. Where the cost of materials handling equipment ten or twenty years ago would have equaled the value of the farm, today it amounts only to a modest increase in investment which may be more than offset by greater output. Applications which were out of the question a few years ago are becoming necessities today. The trend will be in this direction in the future. The farm market offers the prospect of greatly increased use of electric power.

Commercial Sales

The commercial classification includes stores, office buildings, shops, filling stations, and the like. Lighting is one of the main uses of electricity for these customers. Usually the levels of illumination in the stores and offices are higher than they are in the homes. Good lighting helps to sell merchandise and contributes to higher efficiency in the offices. In food stores and restaurants refrigeration is an important load. Restaurants find that electricity has many advantages for cooking and for heating.

Air conditioning has recently come into wide use in commercial establishments, and considerable interest is being shown in electric heating.

The power company usually maintains a staff of people who regularly call upon commercial customers to acquaint them with good lighting practices, decorative lighting, air conditioning, cooking, and heating.

Selling to the Industrial Class

Industrial customers are only one-half of 1 percent of total customers, but they use 43 percent of all the kilowatt-hours sold by the company and provide 25 percent of the company's revenue.

There are several types of power available to industrial concerns. They could use oil engines, gas engines, or steam engines. These engines could be applied directly in the industrial process or they could be used to generate electricity to be used in the plant. However, most industries have found that it is cheaper and more reliable to purchase electric power from the company supplying the area. Usually the cost is less than the cost of other forms of power. The electric utility company is able to make the power cheaper because of the larger, more efficient equipment used by the power company, because of diversity, and because of the many interconnections.

The cost of electricity is one of the smallest items in the total cost of finished manufactured products, as illustrated in Table 7.2.

TABLE 7.2   Electricity in U.S. Manufacturing Industries (1963)
Cost of Purchased Power as a Percent of Product Value

Industry	Percent
Paper and Allied Products	2.2
Chemical and Allied Products	2.0
Primary Metal Industries	1.9
Stone, Clay and Glass Products	1.5
Rubber Products	1.0
Lumber and Wood Products	0.9
Textile Mill Products	0.9
Petroleum and Coal Products	0.8
Fabricated Metal Products	0.6
Electrical Machinery	0.5
Food and Kindred Products	0.5
Furniture and Fixtures	0.5
Instruments and Related Products	0.5
Machinery (except Electrical)	0.5
Miscellaneous Manufactures	0.5
Leather and Leather Products	0.4
Printing and Publishing Industries	0.4
Transportation Equipment	0.4
Apparel and Related Products	0.3
Tobacco Manufactures	0.1
All Manufacturing	0.86

However, this does not mean that the electric utility company does not have to maintain power sales engineers to call upon the industrial customers. On the contrary, the electric utility company considers these contacts quite important. The power sales engineer's job is to help the customer realize the maximum benefits from the use of electric service. The power sales engineer may suggest ways to reduce losses within the plant. He can suggest further mechanization and electrification to reduce costs. He may point out how load factor may be improved in order that the customer may earn a lower rate.   The power engineer is constantly checking to be sure that the service is adequate and the customer is satisfied.

When new uses of electricity are developed, the electric utility company's power sales engineers acquaint their customers with these new developments.

Objectives of Selling

Utility sales efforts are directed toward showing the customer how to get the most value and use from his electric service.   Electricity is energy, and electric appliances are devices which enable the customer to get this energy to work for him.   The power companies show the customer the many ways electricity can serve him, and the rest is up to the customer.

It is true that utilities generally are happy to see increases in kilowatt-hour use and therefore welcome added appliance load.   But the plain fact is that people will not use electrical appliances unless they are convinced that the appliance will do a job they want done at a reasonable price.   It would be shortsighted indeed to sell a customer devices for which he has little need.   The company is far better off if it can sell the customer an appliance that will be used—in other words, one which will serve the customer well.

The refrigerator-freezer consumes more electricity than the old-fashioned refrigerator, but it enables the housewife to store foods in an easy and efficient manner.   She can buy in bulk to take advantage of sales, and need make fewer trips to the store.   Home canning used to be quite a chore; home freezing is simple.   The American housewife has eyed these benefits and decided they are worth the price.   While the companies benefit from this, the customer is getting value received or there would be no sale.

Selling to Increase Percent Return

As mentioned before, the over-all index of electric utility operations is the percent return on total investment. Earnings are not necessarily measured by the gross volume of business or by the net revenue. Net revenue may be increasing in dollar amount, but investment is going up too. If the operating income is not increasing enough to pay an adequate return on the increased investment, then the utility will show a decrease in over-all percent return.

During some periods percent return may be declining. Increases in the uncontrollable expenses may have more than offset the saving brought about by improved mechanical efficiency. The Edison Power Company has plans to improve its percent return. There are numerous ways for a company to improve its percent return. Edison Power Company's management tries to use all of them.

Perhaps the most obvious way is by holding down operating expenses, and that is why all expenses are continually being watched and analyzed. When the company is able to save one dollar in expenses this dollar adds directly to earnings. The company must, of course, pay Federal income tax on that income; but the remainder can go toward improving the return on the investment. Federal income taxes are about 48 percent of net income. Therefore, each dollar saved will yield 52 cents after taxes.

Another way to increase percent return is through sales. However, increased sales means an accompanying increase in investment. This is illustrated in Chart 7.18. Remember that on the average Edison Power Company has invested $4.50 for each $1 of annual revenue. Assume that the company does better than that, and that it takes only $4.00 of new investment to serve the additional business. The company must pay return, depreciation, and taxes on that new in-

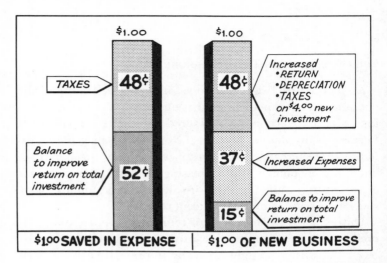

CHART 7.18

vestment. The company must generate more kilowatt-hours to serve the $1 of new business, so there will be increased production costs. The company has a sales expense. There will not be much of an increase in customer accounts expense, distribution expense, or in general expense, although there will be some. Edison Power's economists estimate that the added expense will take 37 cents out of the dollar. By adding up all of these additional costs in connection with the $1 of new business, it is found that there is 15 cents left after taxes. This is the amount available to improve the company's over-all position.

In its sales program the company aims to add new business which will earn a percent return above the company average, thereby increasing the over-all rate of return.

Relationship Between System Annual Load Factor and Percent Return. A good way to improve percent return is by improving load factor. Obviously the company cannot dictate to its customers how they may use the service; the customers use it to suit their own habits and needs. But there are some things the company is doing to improve load factor.

For example, added air conditioning load causes a drop in annual load factor where the system has a summer peak. It helps load factor in such a case when the company induces people to use more electricity during the wintertime. The company has encouraged winter use by means of a vigorous sales program in the house heating field.

The freezer has a group load factor of about 85 percent. The sale of such appliances results directly in the improvement of load factor. The company has conducted several aggressive campaigns on the sale of freezers. Many homes do not now have freezers; consequently, there is a good market for this appliance.

The summer peak usually occurs in the daytime. Any use at all at other times of the day is off peak and tends to help load factor. Most lighting is used in the evening and for that reason is especially good in summer peak areas for raising load factor. But in winter peak areas, it is also a good builder of percent return.

Edison Power Company's engineers and economists have found that the combination of the range and water heater has a much higher load factor than the range alone (Chart 7.19).

CHART 7.19

The RANGE and the WATER HEATER

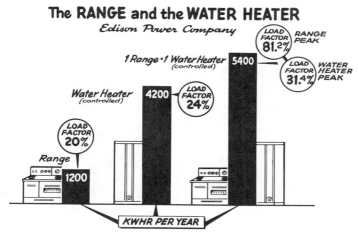

Edison Power Company

Range — LOAD FACTOR 20% — 1200

Water Heater (controlled) — 4200 — LOAD FACTOR 24%

1 Range · 1 Water Heater (controlled) — 5400

LOAD FACTOR 81.2% RANGE PEAK

LOAD FACTOR 31.4% WATER HEATER PEAK

KWHR PER YEAR

This is especially so in the case of the off-peak or controlled water heater. (It is described in Chapter 8. Such a water heater has a time control on it that allows it to run only during off-peak hours.)

Note that the range has a diversified load factor of 20 percent, whereas the controlled water heater by itself has a load factor of 24 percent. The range and controlled water heater together have a load factor of 31.4 percent, considering the peak of the two as the water heater peak. Considering the range as the peak, the combination of the two has a load factor of 81.2 percent. In determining which peak to use, consideration must be given to how the range and water heater fit in with the whole residential load pattern and with the whole system load pattern.

As has been shown, selling controlled water heaters to homes already having an electric range is a good way to increase load factor.

While off-peak control of the water heater results in lower power-plant investment, it usually requires an extra meter, a time switch, and a more expensive water heater. Consequently, many companies prefer to sell the water heater service on an uncontrolled basis.

Interruptible or Off-peak Service. Chart 7.20 shows the load

CHART 7.20

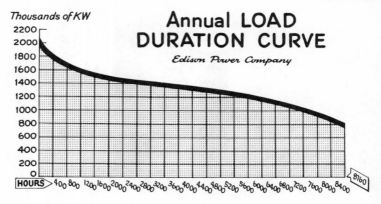

Annual LOAD
DURATION CURVE
Edison Power Company

Thousands of KW

duration curve. The load duration curve shows the distribution of the company's system loads by hours during the year. For example, Edison Power Company's load was above 1.8 million kilowatts for 400 of the 8,760 hours in the year. The load was above 1.4 million kilowatts for 3,000 hours.

The load duration curve tells how long the peak condition is likely to last. Notice the sharp peak in the curve on the left-hand side of the chart. The power company must build and pay for over 200,000 kilowatts to serve that peak condition only during those 400 hours or a portion of it.

If the company could sell this capacity during more hours of the year by improving annual load factor, its earnings would be improved. In some cases a lower rate can be offered to a customer if he limits his demand during system peak periods and thereby reduces the amount of capacity required for these relatively short periods. (Usually when a company sells interruptible service, the hours of interruption are somewhat longer than the theoretical number required. In this way the company reserves some leeway in its operations. Also the company does not give a discount equivalent to the whole fixed charge on the investment, for the reason that, as a practical matter, some investment is reserved for the customer.)

A similar service is off-peak service, which some companies make available to industries which operate only during the off-peak hours and benefit by a lower rate.

The interruptible and off-peak services described here are practical only to a limited extent. As the cost of electricity is only a small fraction of the total cost of manufactured products (see page 181) most customers prefer to buy uninterrupted service.

These are examples of the things Edison Power Company is doing through selective selling to improve its load factor as a means of raising its percent return.

Because of the importance of load factor, the electric power industry carries on a great deal of load research. A representative of Edison Power Company is on an industry committee that works on load research problems. The company makes use of this research and does research of its own in designing a sales program. Knowledge of the load characteristics of all appliances and of the principal uses of service enables the company to use its sales effort to get the highest load factor possible.

Sales Allies. Electric utility companies have many allies helping to promote the sale of electric appliances and equipment. The manufacturers of electrical appliances carry on their own selling efforts. Their salesmen call on dealers, power companies, and distributors; they conduct nationwide campaigns through national advertising media such as magazines, radio, and television. Many manufacturers and dealers advertise in newspapers. There is a national promotion of the whole idea of electric living.

In some cases, the electric utility company itself may sell appliances. The power company's primary interest is in increasing the sale of appliances to the public, whether through the company or through the dealer. Power companies cooperate fully with the electric appliance dealers and try to assist them in making sales. In many cases the power companies concentrate on building up a market for new appliances that do not yet have complete public acceptance, while dealers concentrate on selling the appliances already having public acceptance.

The Design of a Sales Program to Build Volume and Percent Return—Residential Service. In laying out a sound sales program, it is first necessary to analyze existing sales practices and determine what the company's position will be, say, ten years hence, if these policies are continued. Then experts will study various programs designed to better the net results as measured in percent return on total investment.

To find out what the company's growth will be in the future if it follows the pattern of the past, Edison's vice president in charge of sales asked the company's market research department to make projections such as the one shown in

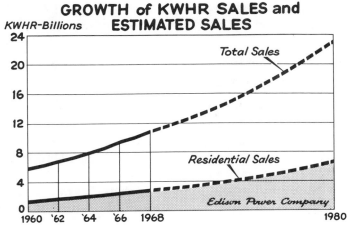

GROWTH of KWHR SALES and ESTIMATED SALES

CHART 7.21

Chart 7.21, which covers total sales and residential sales. Note that the solid line records the performance of the company in the past; the broken line carries this course into the future.

In similar fashion, Chart 7.22 shows the trend in annual load factor of Edison Power Company from 1950 to 1965.

CHART 7.22

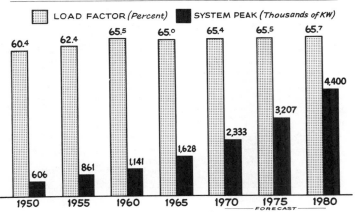

Load factor has been projected to 1970, 1975, and 1980. The system peak load is also shown on this chart for the same years.

Effect of Appliance Saturation

The total load of the residential class of service is a composite of the loads of the various household appliances and other domestic uses. Chart 7.23 shows how these various appliances contribute to a typical day's use of electricity in the winter. The general aim of a sales program is to put greatest emphasis on the sale of those appliances which operate over many hours of the day and on those operating during the off-peak hours.

Since the load curves of different companies will be different, one appliance that looks good should not be arbitrarily seized upon for promotion. For example, some companies in the northwestern part of the country eagerly seek air conditioning sales, because the use of air conditioning tends to fill in valleys in their load curve and improve system annual load factor. On the other hand, in the Southwest air conditioning

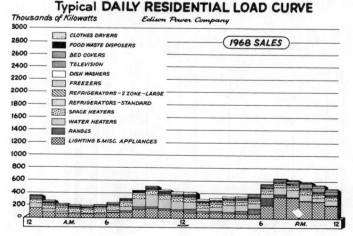

CHART 7.23

Typical **DAILY RESIDENTIAL LOAD CURVE**

Thousands of Kilowatts   Edison Power Company

1968 SALES

CLOTHES DRYERS
FOOD WASTE DISPOSERS
BED COVERS
TELEVISION
DISH WASHERS
FREEZERS
REFRIGERATORS - 2 ZONE - LARGE
REFRIGERATORS - STANDARD
SPACE HEATERS
WATER HEATERS
RANGES
LIGHTING & MISC. APPLIANCES

is directly on peak. In that area, every air conditioner sold tends to reduce annual load factor and thus possibly put the power company in a worse position.

It is necessary to study the individual load curve in each case. It is also necessary to study the saturation and potential market for each appliance under consideration. It is obvious that if two appliances would be equally good for improving the load and load factor, all other things being equal, there would be the greatest potential market for the one with the lower saturation. If appliance X is in use in 99 percent of the homes, and appliance Y is used in only 11 percent of the homes, money spent in promoting appliance Y is likely to produce more kilowatt-hours than money spent in promoting appliance X as well as possibly improving the load factor and the percent return on total investment. This, of course, assumes that each of the appliances has an equal appeal to the user.

Saturation Trends

Past records are examined to find the saturations in each of the major appliances for the past five or ten years. Some companies have these saturation records, but in other cases they can be estimated by checking on the number of appliances sold in the years past. Then these can be compared with manufacturers' sales figures for the area.

For example, Chart 7.24 shows the trend in water heater saturations. The past records, together with the figures from an actual field survey, show the direction in which this line is moving. This can be carried forward to 1980, as shown by the broken line. In like manner, Chart 7.25 shows the trend of dryer saturations with the estimate to 1980. This process is followed for each of the appliances under study.

This part of the study permits a determination of Edison Power Company's appliance saturations twelve years hence under the existing sales program. These saturations are

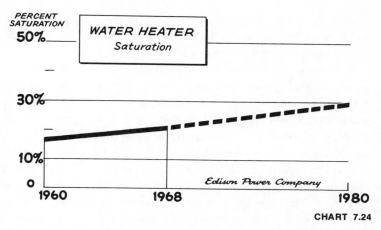

PERCENT
SATURATION

WATER HEATER
Saturation

Edison Power Company

CHART 7.24

shown in Table 7.3. It is known generally how each of these
appliances contributes to the load curve. Chart 7.26 shows
the residential load curve for 1980 based on these estimated
saturations. Note especially the portion of the chart devoted
to space heating.

While appliances will produce the major part of the residential load, lighting will play an increasing part. It appears
that the lighting used by the average customer of Edison
Power Company may increase from 1,100 kilowatt-hours in

CHART 7.25

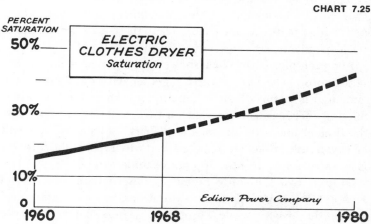

PERCENT
SATURATION

**ELECTRIC
CLOTHES DRYER**
Saturation

Edison Power Company

Typical DAILY RESIDENTIAL LOAD CURVE

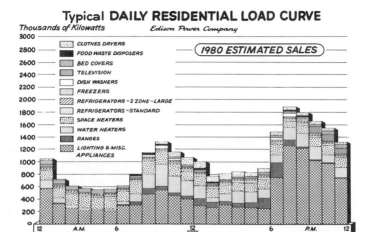

Thousands of Kilowatts   *Edison Power Company*

CHART 7.26

TABLE 7.3   Future Saturations, Existing Sales Program

Appliance	Percent
Water heaters.	30
Dishwashers	35
Ranges.	58
Freezers.	30
Clothes dryers	44
Air conditioners (room).	54

1968 under the normal sales program to about 2,000 kilo-watt-hours in 1980.

Residential Load Factor

These figures and the load curves (Charts 7.23 and 7.26) give enough data so that the annual residential load factor for 1968 and 1980 can be calculated. In this twelve-year period residential sales will increase 161 percent, but the load factor will decline from 48 percent in 1968 to 46.9 percent in 1980, if the company continues its present sales program.

Phantom Appliances

Since new home appliances have been coming onto the market at the average rate of 1½ a year, the company will be furnishing electricity to operate almost 20 new appliances not even known about today. Just what they will look like, how they will operate, or what purpose they will serve no one can say. Nevertheless, they will contribute substantially to growth in kilowatt-hour sales.

Naturally the load characteristics of the new appliances are not known. From the records of the past, it can be assumed that these 20 new appliances will use around 1,000 kilowatt-hours per year. Each company makes the best estimate it can as to where this use will fall during the day, month, or year, and how it will affect its peak load. When actual experience at some future time gives reason to change the estimate, the company's sales program might be reexamined.

While the next 12 years may not produce another frost-free refrigerator, the kilowatt-hour use equivalent to that of another frost-free refrigerator will likely be added to the lines of the company, in addition to the load added by the predictable appliance growth.

These assumptions bring out the picture of a phantom appliance with certain use characteristics. This phantom and its estimated use characteristics should be considered as a refinement in the forecast.

Target Sales Program

The sales department of Edison Power Company has designed a sales program better than the present one. They call this new sales program the target program. It aims to increase the volume, especially in appliances having high load factors. The program puts special emphasis on the sale of appliances which might be used during the valleys in the load curve, so that the end result will be an improvement instead of a decline in load factor.

Actual experience in the change in an individual's load factor with the installation of certain appliances helps in assessing the value of that appliance to the company. Also, it is possible to calculate the probable load factor improvement which will be brought about by the sale of the various appliances. These methods can be used to learn which appliances will help build load factor and how much they will help. For most companies, water heaters, dryers, frost-free refrigerators, freezers, lighting, electric bed covers, and, by and large, the heat pump are desirable loads and will help to increase load factor.

But that is not the end of the problem. A calculation can be made as to which appliances will contribute to earnings, but consideration must be given to what the customer wants. This is what determines whether the appliance can be sold. The best man in the company to give advice on the salability of appliances is the sales manager. By experience, he knows about how many appliances his men can sell in an intensive campaign. He knows which ones are hard to sell and which ones are easy. Of course, this will vary from company to company. For Edison Power Company, the accompanying charts (7.27 to 7.29) show what the sales manager thinks he can do in selling certain appliances. He has made similar estimates for each of the appliances in which management is interested.

Taking into consideration both the appliance's contribution to improved load factor and its salability, a sales program was devised that will make the most efficient use of the money expended in sales expense. Chart 7.30 shows the company's estimate of appliance saturations in 1980 under this target sales program as compared with saturations under the present program.

Chart 7.31 shows the residential daily load curve under the target program for the year 1980.

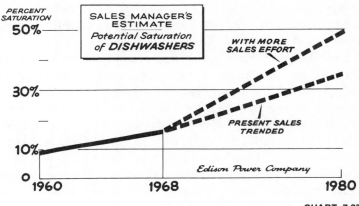

CHART 7.27

With this information, the residential-class load factor for 1980 can be calculated.   This is shown in Chart 7.32.   Note that under the target sales program the load factor does not drop as it did under the normal program.

Increased Sale of the Heat Pump.   Some companies are putting on special programs to stimulate sales of the heat pump. If by doing this the company can sell up to 5 percent satura-

CHART 7.28

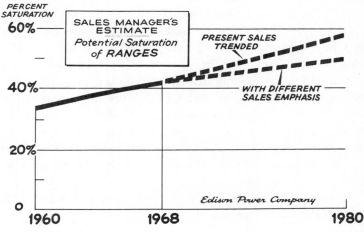

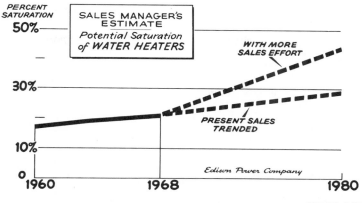

CHART 7.29

tion in heat pumps by 1980, the improvement in the load factor should be as shown in Chart 7.33.

Selective Selling in Other Classes.   The company has followed a similar technique in designing its sales program for the other classes of service—commercial, industrial, as well as the farm market.

Industrial customers are served by power sales engineers

CHART 7.30

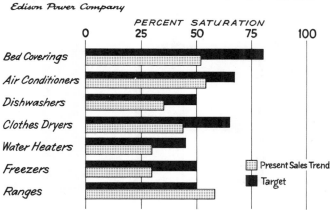

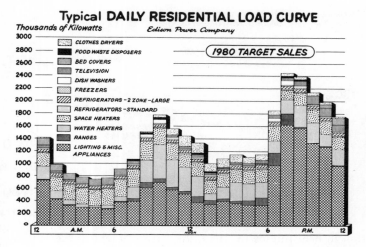

CHART 7.31

who make personal calls on a periodic basis. These power
sales engineers are schooled to help their customers improve
their individual load factors and thereby earn a lower average rate under standard demand-energy schedules.

CHART 7.32

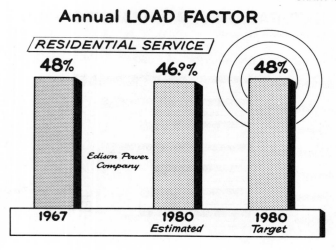

Annual LOAD FACTOR

RESIDENTIAL SERVICE

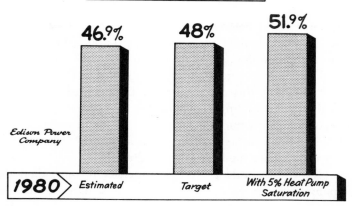

46.9%   48%   51.9%

Edison Power Company

1980 | Estimated | Target | With 5% Heat Pump Saturation

CHART 7.33

Summary

In designing the sales program in this fashion the Edison Power Company analyzes each class of service from the standpoint of building more volume, net revenue, and percent return on total investment. The sales manager watches to be sure that the three classes balance each other out, so that the greatest over-all load factor will be obtained.

To find out how much money the company should spend on a sales program, the gross revenues by classes of service are predicted for five or ten years, and the sales expense needed to get this business is tallied. As might be expected, it is found that the company can afford to spend more money to obtain business that improves load factor. For the year 1968 Edison Power Company spent 1.8 cents per dollar of annual gross revenue in its sales promotion efforts. On a program aimed at raising load factor, cost studies indicate that the company can spend 2 or 2½ percent of its gross revenue in sales effort and still show an over-all improvement in percent return. This is only true, of course, if the rate schedules are at the proper level, especially those rates which cover increased sales.

The Last Resort—Rate Increases. After income is forecast by classes of service under the new sales program, the company makes a forecast for all of its expenses in the manner described in Chapter 6, taking into account all of the local conditions affecting the company and the territory it serves. From this, the company's economists can forecast the percent return on total investment.

If this forecast shows that the trend in percent return is still downward despite the best sales program that can be devised and despite all efforts to reduce expenses further, then the company is faced with the necessity of seeking a rate increase. This has happened to some companies recently, but it is expected that more may face the same prospect in the immediate years ahead, if there is continuing inflation, increased taxes, and high interest rates.

Various possible rate schedules are studied in connection with the long-range new business program. An effort is made to obtain the proper balance between rates and sales. If the rates in some categories are raised too much, a bad effect on sales may result; however, unless the rates cover the cost of incremental service, the percent return will continue to drop even with the best sales program devised.

8

The Nature
of Electrical Loads

The electric power customer's use of energy may vary from
only a few kilowatt-hours a month to sometimes more than a
million kilowatt-hours a month. His demand for electric
service—the greatest amount he will have turned on at one
time—may vary from less than 1 kilowatt of demand in the
case of a small home to 100,000 kilowatts or more in the case
of large industrial customers.

Because of the many variations both in the use of energy
and in the kilowatts of demand, the industry has almost an
infinite variety of customers. Determining the right price for
electricity for the various classes of customers is an intricate
process. Pricing policy—rate making—must weigh all of the
economic factors entering into furnishing service to various

kinds of customers so that all may be treated fairly and so that in the end the company will earn enough net income to enable it to continue as a healthy, going concern.

Here are some of the principles which are observed in rate making:

1. *The rate should be simple and understandable.*

2. *The rate should be salable.* The company is in business to sell electricity. The rate must be fair; but also the customer must know that the price is fair and reasonable or he will not buy the product.

3. *The rate should be competitive.* The electric utility may be the sole supplier of electricity in the area, but that does not mean that there is no competition. Customers may cook or heat water with gas, or electricity. They may heat their homes with raw fuels, or they may heat electrically. Industrial customers would use other forms of power—steam, gas, or internal combustion engines—if they did not consider electricity competitive.

4. *The rate should be promotional.* The rates should encourage the customer to use more electricity. In general, the rate should be designed with a sliding scale, going down with increased use. Where practical, and within limits, the rate should also encourage the customer to operate at a higher load factor.

5. *The rate should be nondiscriminatory.* All customers using service under similar conditions should be billed at substantially the same price. It is necessary and advisable to group customers by classes, but all in the class should be treated substantially the same when the characteristics of service are about the same.

6. *The rate should cover the cost of furnishing the service.* When a company's sale of kilowatt-hours increases, its costs also increase. The rate charged for these additional kilowatt-hours should at least cover their cost. Unless the rates do this, the company's earnings will go down with all in-

creased sales, and the more electricity the company sells, the worse off it will be.

A Matter of Judgment

There is no precise formula for arriving at a rate policy that fits all of these conditions. Management, with the approval of the regulatory body, weighs all the factors involved and then exercises its judgment based on experience to design rates which are in the best interest of the customer and the investor.

Before rate design is discussed, some of the factors that have an effect on the price of electricity will be reviewed.

Load Factor

Load factor has a definite bearing on percent return. Generally speaking, a rise in a company's annual load factor will bring about an increase in percent return, and a drop in load factor will cause a decrease in percent return. Sales to a customer with high load factor have an entirely different effect on the company than sales to a customer with low load factor. Where it is practical to do so, the company's rate schedule will make allowances for the customer's load factor. (The way in which a customer's load factor may be calculated is explained on pages 30 to 34. This explanation also covers system and class load factors.)

Diversity

Diversity may be illustrated by laying out a simple electric system (Chart 8.1). Customers A, B, and C have maximum demands of 5, 10, and 15 kilowatts, respectively. If all three customers use electricity at their maximum demand at the same time, there results a total demand of 30 kilowatts on

ELECTRIC SYSTEM

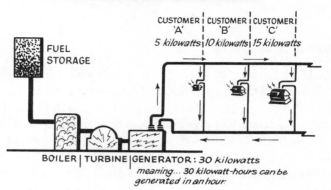

CUSTOMER 'A' — 5 kilowatts | CUSTOMER 'B' — 10 kilowatts | CUSTOMER 'C' — 15 kilowatts

FUEL STORAGE

BOILER | TURBINE | GENERATOR : 30 kilowatts
meaning... 30 kilowatt-hours can be generated in an hour.

CHART 8.1

that system.   The system would need 30 kilowatts of generating capacity to handle this load.

However, the habits of customers vary; each does not make his maximum demand at the same time as his neighbor. During the hour that customer A is demanding 5 kilowatts, customer B may be demanding only 7 kilowatts instead of the 10 kilowatts demanded during the maximum hour.   In that case, the coincident demand of customers A and B is 12 kilowatts (5 kilowatts plus 7 kilowatts).   As these demands occur at the same time they are referred to as *coincident demands.*

As the peak demand of customer A does not occur at the same time as that of customer B, there is a diversity in these two loads.   *Diversity factor,* which is used to measure diversity, is the ratio of the sum of the noncoincident maximum demands of two or more loads to their coincident maximum demand for the same period.   The diversity factor between customers A and B is 1.25 (15 kilowatts divided by 12 kilowatts).

Suppose that customer C uses electricity according to his personal pattern and makes his maximum demand at a different hour from either customer A or customer B.   During the hour A is demanding 5 kilowatts (which happens to be the maximum hour of use on the system), customer C may be using only 11 kilowatts, so that the combined maximum demand of customers A, B, and C is 23 kilowatts (5 kilowatts plus 7 kilowatts plus 11 kilowatts).

This means that the electric system can furnish all of the requirements of customers A, B, and C with a generating capacity of only 23 kilowatts—substantially less than would be needed to serve the total possible demand.

As the habits and needs of individual customers vary, considerable diversity is found in an electric system. The lighting load offers a good example of diversity. There is diversity of lighting load within each house. When the lights are on in one room, they may not be on in the other rooms. There is diversity between homes. When one is making the maximum use of lighting, the next-door neighbor may be out or making little use. The total connected lighting load of all customers in an electric system can be five to ten times the actual coincident lighting demand of all customers on that system.

In addition to diversity among homes there is diversity of use between lighting and other appliances in the home. For example, toasters and percolators are usually used when only a few lights are on. The electric refrigerator runs off and on throughout the day and night. The highest electric range demand does not come at the same time as the maximum lighting demand. Even within the range itself all units may not be creating their maximum demand at the same time.

There is a good deal of diversity among industrial users of electricity. The factory processes being used may be different for each industry. Within each factory there may be a natural diversity in the operation of individual motors.

Adding to these diversities among individual customers, there is further diversity among classes of customers. Residential and commercial lighting customers make their maximum use of energy during the evening hours. Industrial customers make their maximum use of energy during the daylight hours. The combined demand on the electric system is considerably less than the sum of the peak demands of individual classes of customers.

This diversity among classes has the effect of making the combined system load factor higher than individual and class load factors. For example, Edison Power Company's residential service has a load factor of about 48 percent, commercial service about 35 percent, and industrial service about 60 percent. The combined load factor of all these classes together is in the neighborhood of 65 percent. This produces a diversity factor of 1.4 among classes of service. Expressed another way, the coincident demand is 71 percent $(1 \div 1.4)$ of the sum of the separate peak demands. Because of this diversity, the investment in the power pool for the three classes of service together is about 15 percent less than it would be to serve each class separately.

In addition to diversities among customers in one community there is diversity among communities. One community may have a large number of industries; another may be mainly residential and commercial. The habits of residents of one town may differ from those of another. This diversity among towns makes it more economical to supply power from an interconnected system served by a few large power stations than from individual plants in each town.

Additional diversities are realized when one system connects with a neighboring system.

Heat Storage

The controlled water heater makes use of the principle of heat storage. This kind of water heater has a regulator on it that shuts it off during the hours of the greatest demand of the company. The water heater is insulated and stores up heat in the water during the rest of the day. In this way, it uses electricity only when there is relatively little other demand. Such use adds to the total kilowatt-hours used, but does not add to the maximum demand of either the customer or the company. This helps to improve load factor and cuts down on the amount of investment needed to supply electric service. With a lower investment, the price of electricity can be made lower. However, the cost of the water heater is greater and it requires more space.

Research is continuing for a substance capable of storing larger quantities of heat in a more efficient manner. When this is found, customers can apply the principle to house heating by means of the heat pump or direct electric heating.

Load Characteristics

All appliances and equipment that use electricity have certain characteristics in the way they consume electricity. It is important to have a knowledge of these characteristics. Taken as a whole, they form a basis for rate making, for the design of electrical systems to meet possible demands, and for the design of sales programs. Because of their importance, committees of the industry are constantly studying these characteristics. To make the studies, instruments to record the demand at intervals during the day are installed on typical appliances used by typical customers.

The Range. Chart 8.2 shows what a recording demand meter might show for a typical day when installed on a range. This is the load curve of that particular range. Notice that during most of the hours of the day the range is making no demand. During some hours the demand is slight, and during a few hours the demand is heavy. For this particular day the range had a peak demand of less than 3 kilowatts.

Chart 8.3 shows another typical load curve of a range in a home where the customer has his big meal at a different hour of the day. The peak demand of this range occurs at a different time from that of the other customer.

If the curves of these two ranges are put together, the combined load curve will be as shown in Chart 8.4.

Notice that the maximum peak demand is the same for the

Typical Day's Use for ONE RANGE

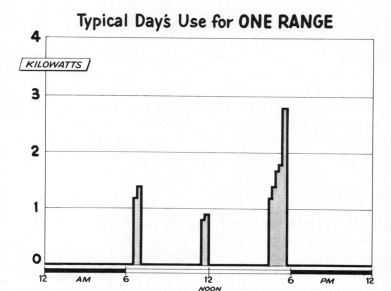

CHART 8.2

CHART 8.3

Typical Day's Use for ANOTHER RANGE

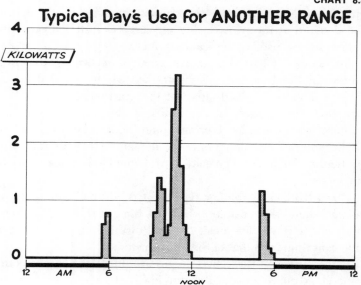

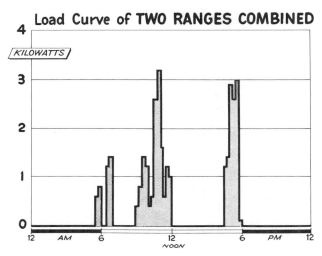

Load Curve of **TWO RANGES COMBINED**

CHART 8.4

two ranges as for the second range. This illustrates the diversity among range users.

Chart 8.5 shows an average load curve for 100 ranges. On the average, the electric range has an annual group load fac-

CHART 8.5

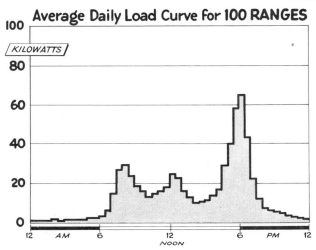

Average Daily Load Curve for **100 RANGES**

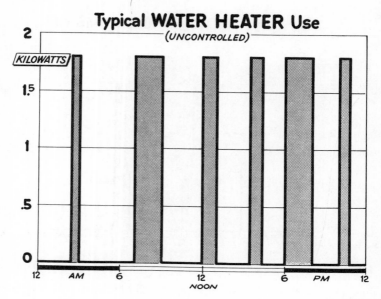

CHART 8.6

CHART 8.7

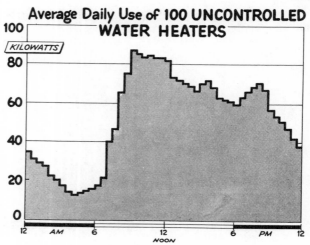

tor of 20 percent. Note that the maximum composite demand of these ranges is 65 kilowatts.

Ranges may have a connected load as high as 10 or even 20 kilowatts, but the diversified demand on the power plant is only about three-fourths of a kilowatt per range.

The Water Heater. Chart 8.6 shows a daily load curve of a typical water heater which has no controls to limit the use during any hours of the day. The water heater is off and on depending on the habits of the customer. Chart 8.7 shows a composite demand of 100 water heaters which are uncontrolled.

Chart 8.8 shows the load curve of a typical range with the load curve of a controlled water heater superimposed on it.

The Refrigerator-freezer. Chart 8.9 shows a typical load curve of a single refrigerator-freezer. Note that it is operating off and on, actually running about 30 percent of the total time.

Chart 8.10 shows a composite load curve of 100 refrigerator-freezers. Notice that the result is a fairly uniform demand throughout the day. This pattern repeats itself fairly

CHART 8.8

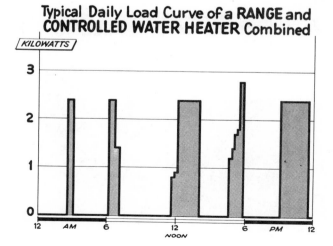

Typical Daily Load Curve of a RANGE and CONTROLLED WATER HEATER Combined

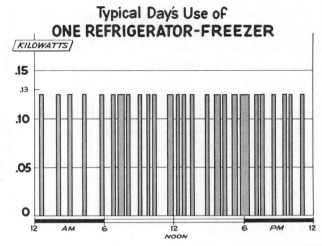

CHART 8.9

uniformly throughout each day so that the refrigerator-freezer on the average has an annual group load factor of about 85 percent.

Lighting and Small Appliances.   Chart 8.11 shows typical load characteristics of lighting and small appliances in a home.

CHART 8.10

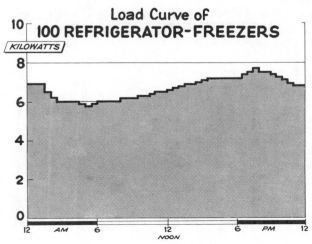

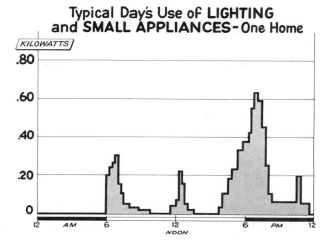

CHART 8.11

Chart 8.12 shows a composite load curve of small appliances and lighting in 100 homes.

The Room Air Conditioner.   Room air conditioners have become quite popular in recent years.   Chart 8.13 shows a typi-

CHART 8.12

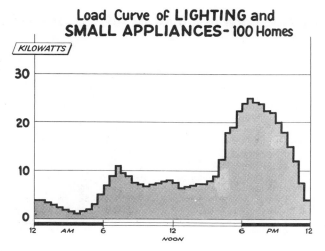

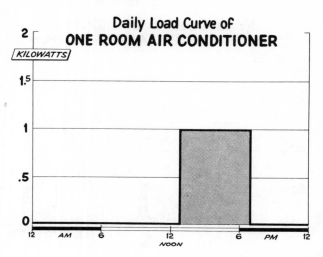

CHART 8.13

cal daily load curve of a room air conditioner during a hot day. Chart 8.14 shows a daily load curve of 100 room air conditioners.

Commercial Load Characteristics. Among the commercial customers lighting is more prominent than in the residential

CHART 8.14

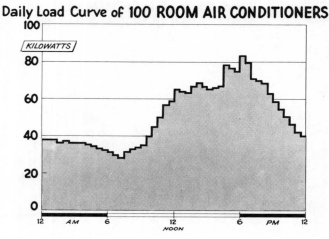

Typical COMMERCIAL Uses

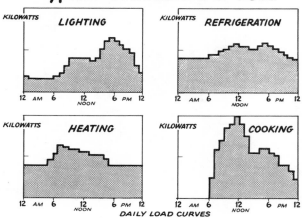

DAILY LOAD CURVES

CHART 8.15

load. Nevertheless commercial customers use air conditioning, some power devices, and some heating devices. Chart 8.15 shows a few daily load curves of typical commercial uses.

Industrial Load Curves. Committees of the industry con-

CHART 8.16

Typical LARGE LIGHT and POWER CUSTOMERS

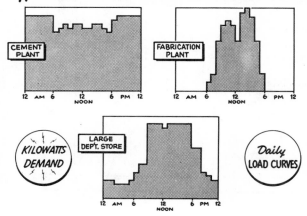

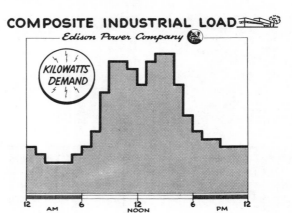

CHART 8.17

CHART 8.18

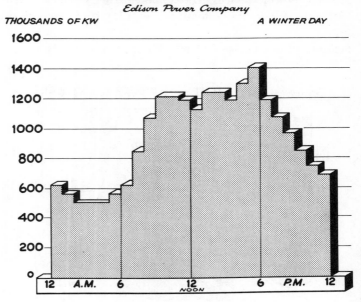

stantly study the load characteristics of industrial and other large light and power customers. Frequently in the case of large industrial customers the company installs recording demand meters as a matter of practice.

Chart 8.16 shows daily load curves of a few typical large light and power customers.

Chart 8.17 shows the composite industrial load curve for the Edison Power Company.

Summary of Load Curve. If the load curves of all customers are combined, the result is a composite system load curve. This is shown in Chart 8.18. This is the composite load curve of Edison Power Company for a typical winter day. Chart 8.19 is for a typical summer day.

CHART 8.19

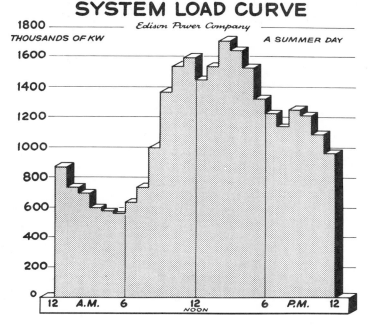

SYSTEM LOAD CURVE
Edison Power Company

THOUSANDS OF KW

A SUMMER DAY

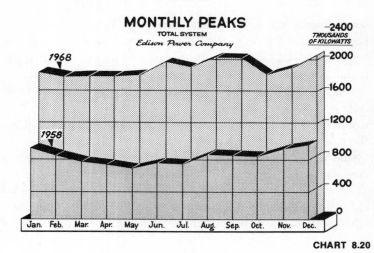

MONTHLY PEAKS
TOTAL SYSTEM
Edison Power Company

CHART 8.20

The top line on Chart 8.20 shows the annual load curve for Edison Power Company. The maximum hour of each month is plotted to produce this curve.

9

Costs and Pricing

There are many factors having a bearing on the price of electricity. The cost of furnishing the service is one of the principal ones. The cost of furnishing service varies mainly in relation to three factors, namely, the quantity used, the load factor at which it is used, and the customer density. The aim in rate making is to set a pricing policy that has a reasonable relationship to cost over the whole range of use and the whole range of load factors. In the interest of simplicity and practicability certain compromises are necessary and advisable.

Variation with Quantity Used

Over the years, the average cost of making electricity has declined fairly steadily. This cost reduction has been possible for two reasons: First, there have been technical improvements, including larger sized generating units, resulting in greater efficiency. Second, today's customers are using more and more electricity, and the principles of mass production have produced still lower costs.

Generally speaking, the more electricity a customer uses, the lower the average cost. For example, the cost per kilowatt-hour of furnishing service to an industrial customer using 1 million kilowatt-hours a month is less than the average cost of furnishing service to a residential customer using 100 kilowatt-hours a month.

Because of this, almost from the beginning electric utilities have followed a pricing policy which provides for a lower average rate for increased use of service. Such a rate schedule is sometimes called a *sliding scale rate.*

The *block rate* in common use today provides for an automatic decline in the average rate for increased use of service. Such a rate may provide, for example, 7 cents per kilowatt-hour for the first 30 kilowatt-hours used per month, 4 cents for the next block of kilowatt-hours per month, and still lower prices for all additional kilowatt-hours used per month. This principle has been incorporated in the rates to practically all classes of customers.

Fixed Costs and Variable Costs

An electric utility company must keep a keen eye on the various elements of cost, so that it can arrive at sound rates which encourage greater use of electricity.

Certain costs are fixed costs; that is, they remain the same whether the business is idle or running at full capacity. Fixed costs can be demonstrated using the cost of an automobile as an example. For example, one pays, say, $3,000 for an automobile, and the installment payments toward this cost are the same each month, regardless of whether the automobile travels 100 or 1,000 miles. These fixed costs may be $100 a month.

Other costs are variable; that is, they go up or down according to how much or how little of a product is made. They compare to the cost of gasoline to run an automobile. If an automobile uses one gallon of gasoline to go 15 miles, and gasoline is 30 cents a gallon, it will cost 2 cents a mile for the gasoline. The total cost of running the automobile will be as shown in Table 9.1.

TABLE 9.1   Cost of Operating Automobile (Example)

Miles driven	100	1,000
Fixed cost.	$100	$100
Variable cost (assuming gas only)		
2¢ a mile	2	20
Total cost per month.	$102	$120
Cost per mile.	$1.02	$0.12

The cost of any article is made up of both fixed and variable costs. Different economic principles apply to situations where, on the one hand, variable costs make up most of the total cost and where, on the other hand, fixed costs are greater. Charts 9.1 and 9.2 illustrate the effect of these two conditions on the price of the product sold.

Chart 9.1 shows costs which are made up mostly of fixed costs. The variable costs are fairly small. This might be the case in a gravity-flow water system where water is free and about the only costs are for the dam and water mains.

The company would have to pay for the dam and mains whether anyone used water or not. To find out the cost of a gallon of water, the company would divide the fixed cost of the dam and mains by the number of gallons. It can be seen that the more gallons delivered, the lower the price per gallon will be. By raising output tenfold, unit costs are reduced 80 percent.

Chart 9.2 depicts costs made up mainly of variable costs. In this case, it is the fixed costs that are relatively small. A good example of this is a commission business where goods are received on consignment and sold for a commission. About the only fixed cost might be warehouse rental. In this case, increased business has very little effect on unit cost. By increasing business tenfold, unit costs are reduced by only 40 percent.

Note that in each of these cases the area showing variable costs is the same width at every point on the chart. The area representing fixed costs diminishes as more units are sold. Note that the downward slope of the top line of the chart is due entirely to the decline in the fixed cost element. Where the fixed costs are relatively high, as in Chart 9.1, the decline in total cost per unit is rapid. Where fixed costs are relatively small, there is a smaller decrease in total cost with increased production.

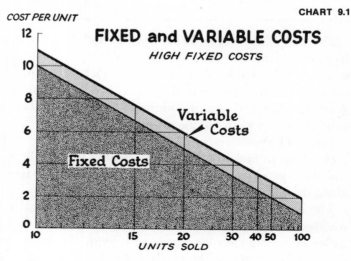

COST PER UNIT

CHART 9.1

FIXED and VARIABLE COSTS

HIGH FIXED COSTS

Variable Costs

Fixed Costs

UNITS SOLD

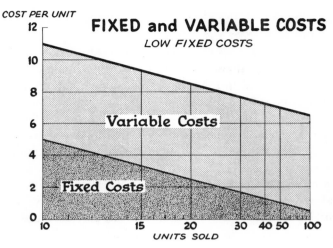

COST PER UNIT

FIXED and VARIABLE COSTS
LOW FIXED COSTS

Variable Costs

Fixed Costs

UNITS SOLD

CHART 9.2

Where the capital investment is high, fixed costs are high. Increasing the amount produced from this investment results in lowering the cost of each item substantially. This is the case in the electric utility industry. A large portion of the income of an electric utility is required to "service" the investment, that is, to pay interest on the money invested, to renew and repair the property, and to pay taxes and insurance on it. These are called fixed capital costs. They remain the same whether the plant operates at 100 percent capacity or is closed down completely.

The variable costs of providing electric service consist, in the main, of fuel, materials, and some maintenance and wages. These vary more or less directly with the plant output and do not decrease on a unit basis with increased output in the same manner as do the fixed costs.

Electric Service Cost Mainly Fixed Cost. About 74 percent of the total costs of providing electric service are fixed costs; the remaining 26 percent are variable costs. This is significant from an economic point of view in that it places the industry in the category of diminishing costs within the limits of an existing plant.

Chart 9.3 illustrates the effect of diminishing costs with increased output. It shows Edison Power Company's cost per kilowatt-hour sold in relation to total sales. Note that its actual sales in 1968 were almost 11 billion kilowatt-hours and that its cost as shown on the chart is 1.12 cents per kilowatt-hour. The chart indicates that if the company could sell more power without installing more generating equipment, the cost would go even lower.

Effect on Cost of Electricity. Chart 9.4 shows the share of fixed and variable costs making up the cost to serve a residential customer. Note that where the customer uses relatively little electricity, the fixed costs are fairly high. As he uses more kilowatt-hours, the fixed cost per kilowatt-hour drops. The variable costs remain the same for each kilowatt-hour. It is the decline of unit fixed costs that gives the curve its downward slope.

This also holds true for commercial and industrial service,

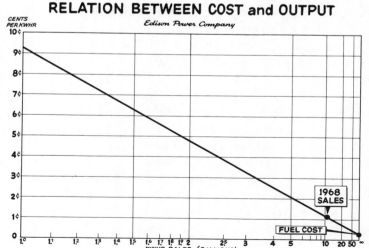

CHART 9.3

RELATION BETWEEN COST and OUTPUT

Edison Power Company

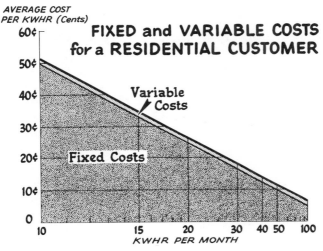

AVERAGE COST
PER KWHR (Cents)

**FIXED and VARIABLE COSTS
for a RESIDENTIAL CUSTOMER**

Variable
Costs

Fixed Costs

KWHR PER MONTH

CHART 9.4

although the cost of furnishing service tends to level off for the larger users. For them, there is not such a great decrease in cost with increased use of service.

The investment needed to serve a customer depends on his demand. If his demand does not increase, the fixed charges on the plant used to serve him do not increase. If he uses more kilowatt-hours at the same demand, then the fixed charges for each kilowatt-hour will go down. In other words, the higher the load factor, the less it costs per kilowatt-hour to furnish service to the customer. Chart 9.5 shows how fixed charges drop with increased load factor.

Power companies take this into account, whenever it is practical to do so, in setting the price for electric service. Rates for larger users provide a discount for customers with high load factor. This encourages the customer to operate at a higher load factor. Also this shows why power companies strive to balance their load to get the maximum over-all load factor. The higher the load factor for the system as a whole, the lower will be the unit cost of furnishing service to the customers.

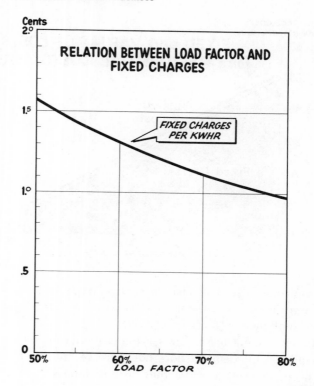

CHART 9.5

Value and Limitations
of Cost Analyses

Any cost analysis requires that the various components of a cost item be sorted out or allocated. It is desirable to know what share of a cost item should be charged to a single customer or to one of the classes of service, for example. The cost analysis may be used to estimate the cost of serving one community or a particular large customer. The analysis might be used to approximate the cost of furnishing service in each of the divisions of a company, all of which obtain their power supply from one common pool. Cost allocations can also be used to determine the way costs of different companies in a pool are to be shared.

Naturally the cost analysis is an estimate and in some measure reflects the judgment of the man making it. The costs allocated to a particular customer or class will vary from time to time as circumstances change with changes of load patterns, changes in characteristics of customers, and changes brought about by growth.

Sometimes the cost analysis is used to figure out the percent return being earned by a company for each of the classes of service. Cost analyses can serve as a guide to assist in maintaining a reasonable balance of earnings by classes of customers. However, the analyses are guides only, as the earnings of the company as a whole are taken into account in determining fair return on the property.

The costs of furnishing electricity depend for the most part on three elements: customers, kilowatt-hours, and kilowatts. All costs can be reasonably classified in relation to these elements.

Customer-related Costs. These are the costs that vary with the number of customers. They include fixed charges on a part of the distribution investment, a part of the distribution operating and maintenance expenses, a large part of accounting and collecting expenses, and some of the expense of sales promotion, as well as a share of general and administrative expenses.

Energy-related Costs (vary with the kilowatt-hours used). These costs consist mainly of a large share of production expenses.

Demand-related Costs (vary with the kilowatts of demand). Primarily these costs grow out of the fixed charges on the power pool and other production and transmission expenses, but are also influenced to some extent by distribution investment and expenses.

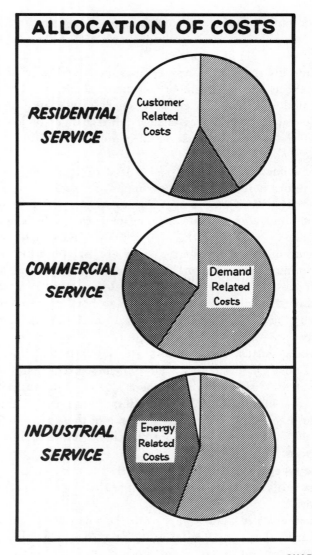

CHART 9.6

Rate Schedules

Most residential and commercial rates are *block rates,* based on the kilowatt-hours used. These provide for an automatic decline in the average rate for increased use of service. Provisions are made in the price per kilowatt-hour to cover all costs including customer-related costs and demand-related costs. In some instances there are minimum bills or per-customer service charges in connection with the block rate. Under the block rate, the rate does not vary with demand.

Other rate schedules include a separate charge to cover the demand-related costs. These demand-energy rates can be better fitted to the cost of furnishing service over the whole range of use.

Few rates make a separate charge to cover customer-related costs.

Chart 9.6 shows how costs are divided for the three main classes of service.

Typical Rate Schedules

The rates referred to herein are used for illustration purposes only. Neither the styles nor the levels pertain to any particular company.

Residential Service. Chart 9.7 shows a residential rate schedule for Edison Power Company. This rate was not designed by any one person as a means of covering the cost of furnishing service for any one moment or any year. Rather, the rate has been developed over a period of years with the judgment of many people entering into its design. It has the required sliding scale or promotional feature. In this form it does not have a demand charge.

This type of rate is appropriate where the range of use falls within certain narrow limits. Until recently, this was true of

Typical RESIDENTIAL RATE
Edison Power Company

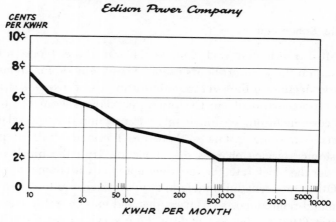

KWHR PER MONTH

CHART 9.7

residential service. Now, however, with new applications of electricity in the home—the heat pump, air conditioning, electric house heating, and other uses—a well-equipped residential customer can use a considerable amount of electricity. For all use above 750 kilowatt-hours, the rate is a flat rate of 1.9 cents per kilowatt-hour. There is no allowance for a customer's load factor. Some companies and commissions have found it desirable to remedy this for the larger user by incorporating a demand principle or a load factor principle in the rate schedule. By so doing, the company can give the customer credit for good load factor. This encourages efficient use of electricity.

Water Heater and House Heating Rates. In some cases a company may add a lower step in the rate to apply where electricity is used for water heating. In other cases there is a lower rate for the number of kilowatt-hours used by the average water heater. Sometimes a separate meter is supplied for off-peak water heaters.

Electric house heating is becoming popular as more and more people come to recognize the many benefits it offers. A company may put a lower step in the rate to encourage the use of either direct panel heating or the heat pump for electric house heating. Some companies put in a separate meter for the house heating, which is charged at a special rate. Companies with a heavy air-conditioning demand in summertime are anxious to sell electricity for house heating to balance out the load.

Importance of the Bottom Step of the Rate. The level of the bottom step in the rate is quite important. The company tries to make the bottom step low to encourage wide use of the service. At the same time it must be high enough at least to cover the added cost of supplying more kilowatt-hours to the same customer. (Without an improvement in load factor, these additional costs include part of the investment in power plants, transmission lines, substations, and distribution systems, as well as the additional operating costs.)

In years gone by, costs in general held reasonably steady or were going down. In those times the cost of making one extra kilowatt-hour—the incremental cost—could be expected to be lower than the average cost of a kilowatt-hour. The more electricity a company made under those conditions, the less it cost.

In recent years costs in general have been rising. Many of the costs of making electricity have gone up. This trend is likely to continue over the foreseeable future. These higher costs change the picture. Companies now attempt to design rates to cover expenses in an economic climate of rising costs. A company may find that the bottom step of its present rate is less than the incremental cost of furnishing service. If that is the case, then the company will lose money on all added sales at that rate. This will hasten the day when the company is required to apply to the commission for a rate increase. Companies and commissions try to avoid this by designing the rate schedule so that added sales will not cause a drop in earnings. A proper rate should maintain the earnings at the proper level.

Residential Rate Trends

Until recently the prevailing trend in electric utility rates was downward. This was made possible by the increased use of electricity, good regulation, and improved efficiency. The resulting savings were passed on to the public through rate reductions.

Even without rate reductions, the average price of electricity will decline as people use more. As most rate schedules are of a sliding scale nature, the more electricity a customer uses, the lower the average price. [If the rate has a "load factor" or "demand" feature, this statement holds for all constant load factors (see Chapter 8).] In that sense the customer's average rate is reduced every time he uses more electricity. The downward slope of the rate schedule controls the amount of that reduction.

Edison Power Company is examining the slopes of these rates in light of the inflationary trends of the past years. It has been noted that both the unit investment costs and unit operating costs have been rising, some rather steeply. The economists of the company are examining the downward slopes of the rate schedules in light of the rising unit costs of doing business. Chart 9.8 shows a few of the rates of Edison Power Company. (For the purposes of this discussion the actual rate levels are not of particular importance. These are not intended to represent the rates of any particular company, nor are they intended to indicate what any particular company's rates should be.)

Chart 9.9 shows the average price of electricity used in the home for the period 1930 through 1968. The trend has been downward in the average price of electricity used in the home.

As shown on the chart, the average price per kilowatt-hour continues to go down even though the rates have been in-

TYPICAL RATES
Edison Power Company

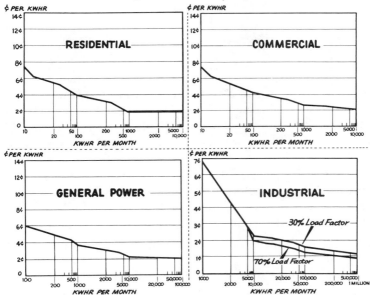

CHART 9.8

CHART 9.9

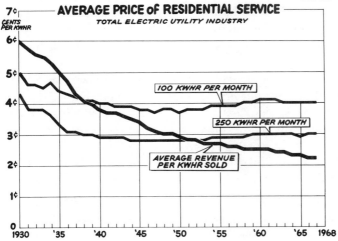

creased. The reduction in average price brought about by the increase in use has more than offset the effect of the higher rates.

The economists working for Edison Power Company predict that operating costs will continue to rise and that there will be more inflation in the years ahead. They question whether the company can continue to hold the line on prices when its costs are rising. This does not mean that the sliding scale feature of the rates should be abandoned. The question is, should the rates go down as steeply as they do or should the slope downward be more gradual? They are considering this.

Chart 9.10 shows the trend in the average cost of making and delivering a kilowatt-hour since 1902. Line A on this chart shows the cost per kilowatt-hour sold with all costs included except the balance for return on investment. In other words, line A includes all operating expenses, depreciation, and taxes.

In line B there is added to line A expenses equivalent to a flat 6 percent rate on investment. This is the total cost per kilowatt-hour of making electricity from 1902 through 1968, if it is assumed that the company earned a 6 percent return on investment. Notice that the trend was generally downward to about 1943, when it rose until about 1954. The trend from 1954 through 1961 was about level. Since 1961 it has again been downward.

However, the actual percent return on investment has not been 6 percent but has been running somewhat less and until recently has been declining.[1] The economists of the company interpret this as meaning that rates had been sliding down too steeply for existing economic conditions.

This is illustrated in another way in Chart 9.11. This chart covers the portion of Chart 9.10 enclosed with dashed lines. Line B is shown, as well as line C, which is the actual reve-

[1] The use of 6 percent here is for illustrative purposes only. The cost of money is now considerably higher.

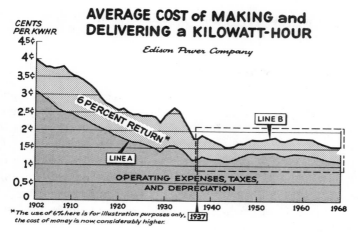

AVERAGE COST of MAKING and
DELIVERING a KILOWATT-HOUR

Edison Power Company

CHART 9.10

nue per kilowatt-hour received by the company at the return
actually realized.

The company's economists have been keeping a careful eye
on this trend, and are reexamining the slope as well as the
level of the company's rates in light of it. They are trying to
achieve the correct balance: rates that would offer a decrease

CHART 9.11

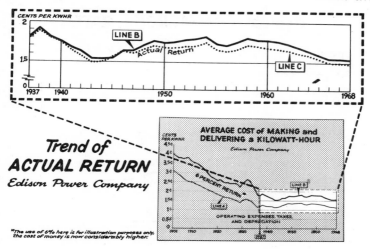

Trend of
ACTUAL RETURN
Edison Power Company

in price with increased use but will not cause a drop in the company's percent return when sales are increased.

Commercial Service

Chart 9.12 shows a typical rate for commercial service. This is the kind normally used for the smaller stores and smaller commercial establishments. It is frequently a simple block rate similar in form to the residential rate schedule. It has longer steps than the residential rate because the use of kilowatt-hours in this class is somewhat higher and the demands are higher. It is especially necessary to have longer blocks in this rate when no allowance is made in the rate schedule for the amount of demand.

This rate also has the sliding scale feature so that the customer may earn a lower average rate with increased use. As in the residential rate, the bottom step should cover the company's incremental cost.

In some companies the length of the first few steps may vary with the kilowatts of demand. This type of rate is illustrated in Chart 9.13. This chart shows the rate schedule for

CHART 9.12

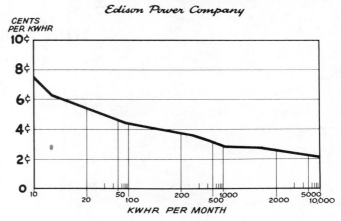

Typical **COMMERCIAL RATE**

Edison Power Company

RATE STEPS VARYING WITH DEMAND

(HYPOTHETICAL EXAMPLE)

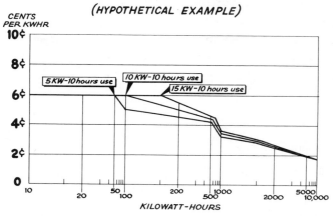

CHART 9.13

5, 10, and 15 kilowatts of demand.   Note that as the demand increases, the steps are longer.   The reason for this is that the higher demands require more capacity and involve more fixed charges.   Under this rate, a small customer at a high load factor can earn a lower rate than a large customer with a poor load factor.   This is proper both from the standpoint of the customer and from the standpoint of the company. For the larger commercial installations, such as hotels, office buildings, and department stores, a demand-energy rate is often used.

Industrial Service

The industrial class includes the large users of electricity, such as manufacturers, refiners, and a host of other industries. Because they use large amounts of power, it is important to have their rate fit the nature of their use and the cost of serving them.   The industrial rate takes into consideration both the number of kilowatt-hours used (the energy charge) and the kilowatts of demand they impose on the system (the demand charge).

The demand-energy rate is designed to cover the fixed costs and the variable costs in all combinations. It has a sliding scale demand charge to which is added a sliding scale energy charge. This type of rate gives a customer his due benefit both for increased use and for higher load factor; the larger the customer, the lower the average rate he earns at the same load factor.

It is possible to design a rate of this kind to cover a very wide range of use. However, some companies prefer to have separate class rates for each type of industry. The tendency is to reduce the number of rates available to simplify the whole rate structure. The trend is toward fewer rate schedules.

The Value of the Service

Electricity has a certain value to the customer. He has a general feeling as to what it is worth to him. If the price is set higher than this he will not use as much. Companies try to set the price so that the customer will want to use it for cooking, for example, and for refrigeration, water heating, clothes drying, dishwashing, house heating, and air conditioning; but at the same time the price should cover the costs.

Competition

Competition plays a greater part in the design of rate schedules than most people realize. The rates to large industries must be competitive with other forms of power, or the customer will use the other forms. He could use diesel power or steam engines to turn his machinery. He may decide to install a generator and make his own electricity. Power companies and commissions constantly study these competitive forms of power so that the price of electricity to industry will be in line.

Competition also comes into play in the other classes of

service. The commercial establishment can use other forms of power. It could install its own generating equipment, but few of them do because, as a rule, the electric rates meet the competition. The larger generating units and the diversity in a large interconnected system enable the power company to sell electricity for less than the cost of power produced by such individual plants. A commercial customer may also use gas for cooking, heating, or air conditioning, or he may use other forms of fuel. It is the electric industry's aim to have the commercial rates meet this competition.

Because of the importance of price, electric utility companies are striving to find better and cheaper ways to make electricity. There are no secrets in the power business. Through various committees in the trade associations, the companies exchange ideas and experiences with each other. When one company learns a better way to do a certain job, all have access to that knowledge. When one company makes a mistake, all the companies can profit by its experience.

Cost of Furnishing Service

While rates should be designed so that each class pays its fair share of the costs, it is not possible or practical to design rates so that all steps in all the rates result in the same percent return on the investment. The aim is to make the prices bear some reasonable relationship to cost. The rate should not bring an unreasonable benefit or hardship to any particular customer. It should be such that increased sales will result in increased earnings, but at the same time it should encourage the customer to use more electricity.

Taking all of these factors into account, rate making is largely a matter of judgment based on experience.

10

The Demand Charge
and Changing Rates

The demand charge is often not fully understood. Some people think it is an extra charge in addition to their regular bill; others cannot see why the pricing has to be in two parts—one the demand charge, and the other the energy charge.

Part of the confusion about the demand charge stems from the fact that many people do not distinguish between kilowatts and kilowatt-hours. The term "demand charge" itself is unfortunate. It might be better if the term were called the "capacity charge," so that it might be distinguished from the energy charge. The term grew from the concept of the customer making certain demands on the power company system. It signifies the customer's rate of use of energy—the kilowatts of generating capacity the company must reserve for him.

The electric utility company maintains a system of power plants, transmission lines, and power substations. The system has a definite capacity in kilowatts, just as a hotel has a definite capacity in the number of rooms, an office building has a definite capacity in the number of offices, and a water pump has a definite capacity in gallons per hour.

Separate Charges Advisable

In designing a rate to cover all of the fixed and variable costs of rendering the service, three alternatives are possible:

1. All the expenses, both fixed and variable, could be covered by a single charge based on the customer's demand.

2. All the expenses, both fixed and variable, could be covered by a single charge based on the customer's use in kilowatt-hours.

3. The fixed expense could be considered separately from the variable expense. The rate could then be designed in two parts. One part of the rate would be based upon kilowatts of demand; the other would be based upon the kilowatt-hours used by the customer.

In any case, the sliding scale principle is desirable because the unit cost of furnishing service decreases with the increased use of both capacity and energy.

Here is how these three alternatives might work:

Block Capacity Rate (based on kilowatts of capacity required by the customer). The charges to the customer under a block capacity rate could be based on the kilowatts of capacity required. There would be no additional charge for the quantity of energy in kilowatt-hours used. The cost of an average amount of energy would, of course, have to be included in the charge for capacity. This type of rate is sometimes used in serving very large industries, especially in the case of hydro generation. However, it is not widely used.

Block Energy Rate (based on kilowatt-hours only). In the design of this rate, the charge for kilowatt-hours must cover both the fixed costs of furnishing capacity and the variable costs of delivering energy.

The amount of capacity required to deliver the same number of kilowatt-hours to different customers varies over a wide range because some customers use their capacity many hours a day—even twenty-four hours a day in some cases— where others use their capacity only a few hours a day. With no specific charge in the rate to cover the cost of the varying amounts of capacity required, the charges for kilowatt-hours must cover the fixed costs of an average amount of capacity.

There are classes of users whose load factors naturally fall within such definite limits that a rate of this type can be designed to fit the average load factor of the class. While it will not do exact justice to each individual member of the class, the benefit of simplification may more than offset the slight discrepancies that occur. This is true in the case of domestic service for which this type of rate is now almost universally used. It is also true for a large proportion of small power users with whom it is common practice to use this block energy rate in combination with an optional rate of the demand and energy type.

In the case of larger commercial and power customers load factors vary between wide limits. The following example shows how inequalities can result both to the customer and to the company from a rate based on energy only. Assume there are two customers, each using 1 million kilowatt-hours a year. The Jones Company requires 200 kilowatts of capacity (200 kilowatts demand); the Smith Corporation requires 500 kilowatts of capacity (500 kilowatts demand).

Consider only two primary costs of furnishing service, namely, fixed charge on the investment and production cost. Table 10.1 shows the difference.

Other costs, not included in this example, would make the total cost higher than shown. Both customers use 1 million

TABLE 10.1   Cost Comparison

Cost factor	Jones Co.	Smith Corp.
Kilowatt-hours per year	1,000,000	1,000,000
Kilowatt demand	200	500
Average monthly load factor	57.1%	22.8%
Coincidence factor	0.82%	0.75%
Kilowatt demand at plant	164	375
Investment @ $240 per kilowatt	$39,360	$90,000
Fixed charges 17% (cost of money 7%, depreciation 3%, taxes 6%, insurance 1%)	$6,691	$15,300
Production cost @ 3.75 mills per kilowatt-hour	$3,750	$3,750
Total fixed charges plus production cost	$10,441	$19,050
Cost per kilowatt-hour	1.04¢	1.91¢

kilowatt-hours, yet the company's cost of furnishing service to the Jones Company is considerably lower than the cost of serving the Smith Corporation. If each of these customers is charged on the basis of the kilowatt-hours used without any reference to the kilowatt demand of each, both would pay exactly the same price because both used the same kilowatt-hours. Either the Jones Company would be charged more than it should be, or the Smith Corporation would be charged less than the cost of serving it.

The only way to avoid these variations is to have a two-part rate with both demand and energy charges. The result is a rate that is in line with the cost of furnishing service under varying load conditions. That rate will also compete with the customer's cost of running his own plant. The two-part rate does not mean that the customer is being charged twice. It merely divides the total cost of serving him into two parts.

The Two-part Rate

The two-part rate has a sliding scale demand charge to which is added a sliding scale energy charge. Following is a sample, although it pertains to no particular company or area:

Demand charge

$2.25 per month per kilowatt for the first 50 kilowatts of the maximum demand in the month.

$2.00 per month per kilowatt for the next 150 kilowatts of the maximum demand in the month.

$1.75 per month per kilowatt for all additional kilowatts.

Plus an energy charge

3.0 cents per kilowatt-hour for the first 1,000 kilowatt-hours used per month.

2.0 cents per kilowatt-hour for the next 2,000 kilowatt-hours used per month.

1.0 cent per kilowatt-hour for the next 7,000 kilowatt-hours used per month.

0.8 cent per kilowatt-hour for the next 90,000 kilowatt-hours used per month.

0.7 cent per kilowatt-hour for all additional kilowatt-hours used per month.

The following shows the average rate per kilowatt-hour that Jones Company and Smith Corporation would earn under this demand-energy schedule.

	Jones Co.	Smith Corp.
Average rate earned, demand-energy rate	1.37 cents per kwhr	2.0 cents per kwhr

The rate, of course, covers costs not included in the example on page 243.

Demand Charge for a Customer's Own Private Plant

Suppose a customer decides to install his own plant. If he needs 500 kilowatts of capacity, he must purchase a machine that will deliver this capacity. (Reserve capacity must also be provided, if the service is to be even partially comparable with electric utility company service.) The customer must invest a certain amount of money in the machine. On this investment he must pay interest, depreciation, taxes, and insurance. Other items such as labor and maintenance are also fairly well fixed.

Assume, for example, that the customer must invest $250 a kilowatt (including reserve capacity) in his own plant. Assume that the cost of money is 7 percent, depreciation is 7 percent, and taxes and insurance amount to 5 percent, or a total of 19 percent. This amounts to $47.50 per kilowatt ($0.19 \times 250) a year, or $3.96 per month per kilowatt of capacity.

These fixed charges are similar to the demand charge in the electric rate. In other words, the customer must pay a demand charge for his capacity requirements whether he purchases capacity outright in the form of an engine or whether he rents that capacity from the electric company.

Parallel Cases. Demand charges are neither uncommon nor unusual. They are met under a different name every day in many businesses. Rental charges for office space are demand charges. When a tenant rents three offices in a building on a monthly basis, he expects to have three offices reserved for him and available every day and night for the period of the rental. The tenant may use these offices one day a month or thirty days a month, but he expects to have them there available to him at all times.

If he uses the offices only a few days a month, his "occupancy factor" is low, and his cost per day of occupancy is greater than if he were to occupy the office every day.

If the tenant occupies the office for three days during any one month and divides the total monthly bill by the number of days' use, he will find a rather high charge per day. He will be willing to pay this charge because he knows the offices have been reserved for his use. In the contract with the building owner, he demanded that the offices be kept available for his use whether he used them or not. The renter of capacity in a building can visualize that the space is reserved, whereas the electric customer has difficulty in visualizing the reservation of the space or capacity that he is renting in the electric utility system. The two are, however, analogous.

The same analogy would hold in the case of a tenant renting rooms in a hotel on a monthly or annual basis. If that hotel tenant wants meals, he would expect to pay an additional charge for each meal. This charge would be added to his rental charge for space in the hotel. This is the case of a demand charge plus a commodity charge. He need not pay for the meals unless he actually consumes them. The electric customer need not pay for the energy unless he actually uses it. If the hotel tenant eats only a few meals a month at the hotel and then divides his total monthly bill for room and meals by the number of meals purchased, the result will be a ridiculously high price per meal.

The automobile storage company charges so much per month for each stall. The customer pays $35 a month for that stall whether he puts the car in the garage one night or thirty nights a month. Similarly, if he rents space for twenty-five cars on a monthly basis and requires the building owner to have this space available for him at all times, he must pay accordingly.

Suppose a man calls the taxicab company and wants to buy 100 passenger-miles of travel. The rate that he should pay per passenger-mile depends upon how many taxicabs he needs to use the 100 passenger-miles. The price per passenger-mile for 100 cabs traveling 1 mile would be higher than the price for one cab traveling 100 miles. In the first case, the customer is demanding 100 times as many cabs and tying up 100 times as much of the company's investment; yet a rate based only on the number of passenger-miles would charge the same amount in the two cases.

Telephone Company. The telephone company is another example of a business that charges on the basis of capacity required. If a businessman wants to use one telephone, he pays, for instance, $9 per month for the use of the telephone, allowing up to seventy-five calls per month. He pays that whether he makes one call or seventy-five local calls a month. If a customer wants to use five telephones, there is a higher price for five telephones. If he wants to use 100 telephones, there is a price for the use of 100 telephones. There is a sliding scale of so much per telephone per month. This is a straight charge without regard for the amount of use.

In effect, the telephone company has a capacity or demand charge to which is added the equivalent of an energy charge. There is a monthly charge of so much per month per telephone to which is added the unit charge of so much per call. The telephone company is like the electric utility companies in many respects. It must meet all demands of its customers (although the telephone company can say, "The circuits are busy," whereas the electric company cannot fail to light the light when the customer throws the switch). The telephone company has a large investment with respect to annual gross revenue.

Changing Rates

Rates are seldom changed radically at any one time, and there are rarely any great changes in the form of rate. The company generally tries to avoid any disrupting change in the billing to individual customers. It is likely to work out a long-range program of standardization, simplification, and improvement in form and make the changes as conditions justify.

Naturally the commission's viewpoints have a bearing on the forms of the rates. The commission's interest in these forms is the same as the company's interest—to have them simple, understandable, salable; to cover the value of the service, and reasonably to cover the costs.

When any rate changes are made, complicated and exhaustive studies are made to find out what effect they will have on the company's revenue. Studies are also made to see how the changes will affect individual customers and classes of customers. It would be impractical and costly to calculate the bill of each individual customer on the new proposed rates. Certain short cuts have been devised which give accurate results. One such short cut is described below.

Frequency Distribution of Kilowatt-hours—Ogive Method. In residential service, for example, a very few customers may use only 1 or 2 kilowatt-hours a month or 10 kilowatt-hours a month. A few will use 1,100 or 1,300 kilowatt-hours or more each month. The great bulk of customers will use something in between. To test the effect of a proposed rate it is necessary to know how many customers are using any given number of kilowatt-hours a month. This will give a picture of how the customers are using the service. Chart 10.1 shows the distribution of customers by their kilowatt-hour use for the residential class of service for Edison Power Company in 1968. This is called a frequency distribution curve. From this curve can be figured the number of kilowatt-hours being sold at each step in the rate. For example, the number of kilowatt-hours a month being used by customers who would fall under the first rate block of, say, 25 kilowatt-hours per month can be read from the chart. To see the effect of a new rate, these kilowatt-hours are multiplied by the price per kilowatt-hour for the first block of the new rate to determine the revenue that would be produced in that block. As a check on

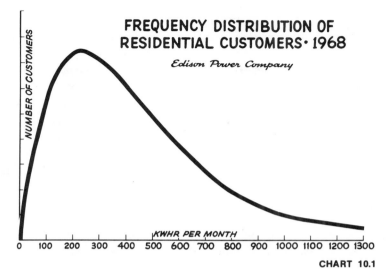

FREQUENCY DISTRIBUTION OF
RESIDENTIAL CUSTOMERS · 1968

Edison Power Company

NUMBER OF CUSTOMERS

KWHR PER MONTH

0   100   200   300   400   500   600   700   800   900   1000 1100 1200 1300

CHART 10.1

this frequency distribution, it is tried out on the existing rate schedule to see if it gives an accurate result.

In the same way, the effect on revenue can be calculated for any rate changes in the commercial or industrial class. Normally the changes applied to large industrial customers must be figured individually.

Distribution Computations Facilitated.   There was a time when the construction of these frequency distributions of customer use meant counting individual customers from meter records. Before data-processing machines were available, this was a very laborious process. Some years ago, however, it was discovered that by the use of ogives and probability paper, families of curves could be constructed empirically from which the same information could be developed as is obtained from the physical count of customers, and with much less effort and time. This method, developed by William Parkerson, is called the *ogive method* and has received broad acceptance among rate analysts in companies and on the staffs of regulatory commissions.

Not only does the ogive method facilitate the construction

of customer distribution curves from current data, but, by estimating future average uses, curves can be constructed which permit the forecasting of future revenues from rates, after taking increased use into consideration.

If a company is planning a new rate schedule, the company can apply the new rate to the frequency distribution curves by years for the next five years or so, in order to find the revenue it will get. Assuming that the forecast of kilowatt-hours is correct, the calculation of revenue can be made with a probable error of less than 1 percent.

Rate Increases

In recent years some companies and commissions have found it necessary to increase rates in order for the companies to keep their earnings at a satisfactory level to maintain good credit and a sound financial position. When rates are increased, new rates are designed with the aim of maintaining the company's earning position for a number of years.

By the method described above it is possible to try out new rate schedules to see what the company's earnings might be. Both the commission and the company would prefer to avoid the expense of periodic rate cases. They both prefer to design the rates so that they will maintain the company's earnings at a satisfactory level for a period of at least five years.

In a rate case the commission normally considers the actual records of the company for the most recent year. Forecasts might be reviewed but the decision is based upon the actual record.

A forecast is made of the kilowatt-hours, kilowatts, and customers by years and by classes of service. The ogive method shows the usage pattern of the customers by classes of service. These kilowatt-hours are then priced at the proposed schedules for a period of five years to determine the projected gross revenue. Forecasts are then made of all items of expense to arrive at the amount that might be expected for return.   Finally, the percent return on the investment is figured.

By following this procedure, a sound and reasonable rate schedule can be developed for each class of service.

11

The Evolution
of Power Supply

Looking back over the evolution of power supply as it has taken place in this country over the past sixty or seventy years, it is possible to see a number of trends. Knowledge of these trends can be combined with forecasts of what may be coming to give a preview of what may evolve between now and the year 2000. This evolutionary development may be reviewed by periods, allowing students to see something of the problems faced by power industry people and to see how these problems have been met through innovation and change. The attempt of the industry always has been to get electricity to people where they want to use it and to render the service reliably so that they can count on its being as continuous as possible. At the same time, the companies have striven for higher and higher efficiencies in operation through all the avenues available to them in order to bring about low prices which, in turn, would encourage people to use more electric energy.

Power people have been conscious of the fact that the highly refined energy form they supply has an important impact on living standards. It drives the machines people use in their homes and on their jobs. It drives the machines in factories so men can multiply their productivity and produce more goods for themselves and for others.

In the early days of the industry power plants were small and they were isolated. They were built where people were. These small isolated plants were much less efficient and considerably less reliable than the large interconnected systems of today.

As early as 1911, Samuel Insull demonstrated that better and more economical electric service could be provided if a number of small communities having small isolated plants were interconnected with transmission lines and the electric energy brought from one larger and more efficient power plant with satisfactory reserves. His proof was the system he built in Lake County, Illinois. It was, in fact, a research project—and it worked. The country began to witness a growing practice of building transmission lines, of interconnecting power suppliers, of shutting down less efficient power plants, and of replacing them with larger and more efficient ones. All during this period of growth more and more people were able to receive the benefits of electricity. Small towns that could not justify an electric generating plant were served from a transmission line passing nearby, and the process of rural electrification began (Chart 11.1).

The trend toward interconnection had advanced considerably by the start of World War II. All but the most thinly populated areas of the country were interconnected by high voltage transmission lines (Chart 11.2). As interconnections developed and the systems grew in size, the industry was

High-Voltage TRANSMISSION LINES
60,000 VOLTS AND OVER

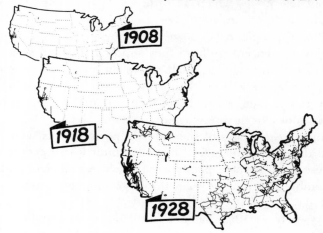

CHART 11.1

CHART 11.2

High-Voltage TRANSMISSION LINES

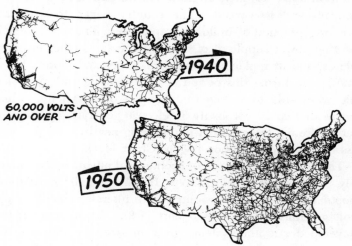

able to make use of larger and more efficient generating units. There are few industries which have been so successful in taking advantage of the economies of scale as advancing technology brought them into reach.

In all this upward sweep of unit sizes, toward higher voltages and increases of interconnections, new problems had to be faced at each step along the way. Almost every new generating plant was designed to be a little larger and, if practical, to operate at a little higher temperature and at a little higher steam pressure. This presented metallurgical problems in the construction of tubing and turbine blades. Each step upward in voltage meant more research, more development work, and more experimentation to identify and solve new problems in switches, breakers, relays, and transformers —all of which had to be redesigned, tested, and proven in practice.

Chart 11.3 shows Edison Power Company's generating plants and transmission lines with former locations of smaller plants. As the company could be organized at a satisfactory size so as to preserve the benefits of good local management as well as to take advantage of the benefits of scale, it was found that it could interconnect with neighboring power systems and bring about further benefits both in reliability and in economy.

Interconnection also helps to improve the load factors of interconnected systems because of diversity between systems. Chart 11.4 shows how Edison Power Company is connected with two other companies to the north and to the west.

Edison Power Company profits through these interconnections. The company has a peak demand of 2,056,000 kilowatts and a total capability of 2,439,000 kilowatts. The value of interconnections showed up strikingly in 1964. Ordinarily, the company would have to have reserve capability of about 20 percent in order to be sure it could carry the load during any kind of emergency. In that year, the company's reserve was on the order of 18 percent. Two of the company's plants, accounting for about 19 percent of total capability, were put out of commission by a spring flood.   As interconnections enabled the company to call upon the facilities of its neighbors for emergency service, Edison Power's 18 percent reserve was enough to see it through, and it was able to furnish its customers a continuous supply of electricity.

The power company maintains reserve capability so that the electricity supply will not be interrupted in case of a breakdown in one or more of the generating units.   Also, the extra capability allows room for growth and possible load forecasting errors (the load that actually develops might be more than was predicted several years before when the sys-

CHART 11.3

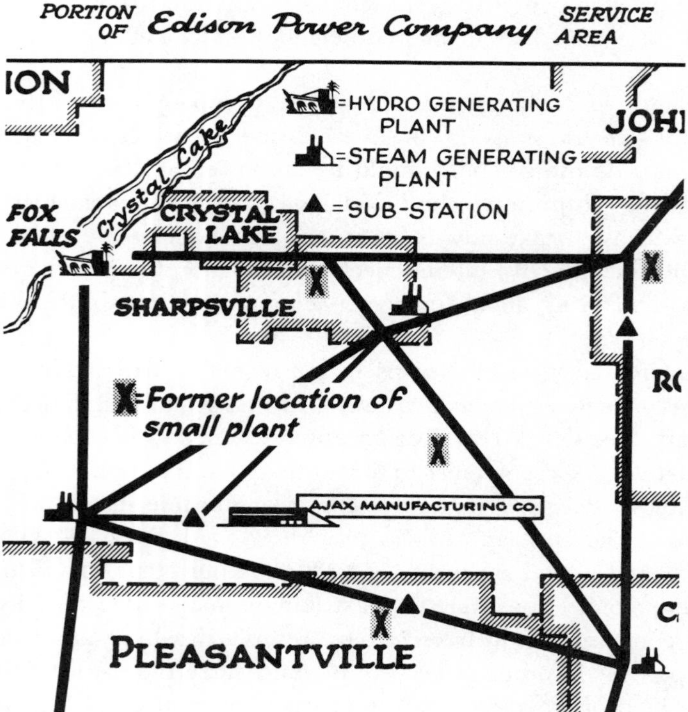

PORTION OF *Edison Power Company* SERVICE AREA

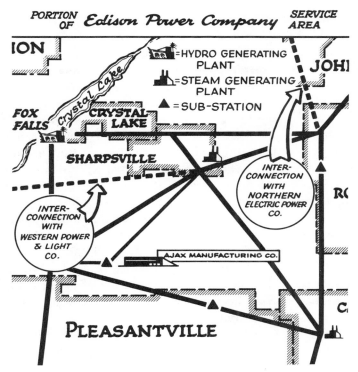

PORTION OF *Edison Power Company* SERVICE AREA

CHART 11.4

tem additions were planned). In some instances, additional
capability is required because it is necessary to take equipment out of service even on peak in view of the fact that the
"valleys" between seasonal peaks are not dependable enough
and long enough to permit the necessary maintenance of
equipment. The company aims to keep the reserve capability as low as is practicable and still maintain dependable service. The reserve capability costs money, because it produces no kilowatt-hours and therefore no revenue. The company aims to keep reserve capability somewhere around 15 or
20 percent of its estimated peak load demand.

Assume that a system has a peak load of 1,700,000 kilowatts

and that it has 20 generators, each of 100,000-kilowatt size to carry this load.[1] Thus it has a reserve capability of 300,000 kilowatts, or approximately 17½ percent of its peak load. This "unused" capability or reserve allows for the contingencies, possible load forecasting error, and maintenance needs.

At any given time, the system must have generating facilities in operation with sufficient capability to pick up any reasonable amount of load instantaneously placed on the system or to replace generating capability that is lost through malfunction or failure. This capability is called "spinning," or more properly *operating reserve*. It must be available almost instantaneously or within a very short period of time.

Thus, in the example cited, while seventeen of the system's 100,000-kilowatt units would have sufficient capability to meet the peak load requirement of 1,700,000 kilowatts, the system will use eighteen of its units to carry the load, possibly with each of them running at 95 percent of their capability, so that in case one unit is lost the others will have sufficient capability to carry the load.

To ensure dependable service, the system must be able to carry its peak load when the largest unit is out. Because of this, a company of average size cannot take advantage of the larger units with their greater efficiency unless it maintains an overly large reserve. The extra cost of carrying a big reserve often offsets the economies of the big unit. Interchange power agreements with neighboring companies can in effect take the place of reserve, in that the interconnecting systems can spread the risk of outage over many units. That is, the simultaneous outage of equipment on each of several interconnected systems will be less probable than the outage on a single system. This is but one of several benefits that can be derived from interconnected operations.

[1] The generating units in the plants of any power company are of varying sizes, as they have been built over a period of years. However, it is easier to show the need for reserve and the value of pooling if a simplified example is used by assuming generating units of the same size.

It is common practice for neighboring companies to contract with one another to interchange electric energy. These contracts may take a number of forms.

1. *Purchased Power.* Frequently one company will purchase power from another company on either a short-term or a long-term basis. The purchase may be for firm power from a system's total resources or it may be from a specific generating unit. In the former case the seller will meet its commitment to the purchaser essentially as if it were part of its system load. That is, it will not only supply the power under normal conditions, but it will also have adequate reserve to assure a continuous supply of power to the purchaser even under adverse conditions. This type of arrangement provides a simple approach for a small system to realize the benefits of economies of scale, lower fuel costs, and lower reserves.

Power companies also contract to sell and purchase power from a specific generating unit. Normally this is done on a short-term basis when power companies wish to cooperate in building a unit larger than either might wish to build alone, although this need not be the only circumstances under which arrangements are made concerning the power output from a specific unit. In this type of arrangement company A might build a large unit and company B might agree to purchase a fixed number of kilowatts from it if and when it is available. Company B furnishes the reserve.

There are many variations of power-purchase contracts, depending on specific needs of the companies involved.

2. *Exchange of Capability.* A variation of the power-purchase contract is one in which neighboring companies might exchange, rather than purchase and sell, capability. That is, power company A will build a large unit and make available a certain portion of the capability to company B for a certain period of time. Then at the expiration of this time company B will make available to company A an equal number of kilowatts. In this way the power companies will plan their sys-

tem expansion together and stagger their building programs. The combined loads of the two enable them to build larger units, taking advantage of the efficiencies to be gained from increased size.

3. *Exchange of Economy Energy.* To keep production costs down, the power company tries to supply its load from the most efficient generating stations as much of the time as possible. The less efficient equipment is operated when needed or kept as reserve. For example, company A may have a generating station which is not fully loaded. If the production cost of the plant is lower than plants of company B, company B may buy power on a temporary basis from the plant of company A. Company A makes no commitments as to firm power but is willing to let company B have energy from this lower cost plant as long as company A does not need the energy itself. Company A makes a small profit on this energy, but the price might still be less than it would cost company B to generate it. This energy is sometimes referred to as *economy energy.* Both companies benefit when between them they can keep the most efficient generating stations running steadily.

4. *Dump Energy.* In some hydro plants there is an abundance of water because of continuous stream flow. Since the production cost of a hydro plant is low, the company naturally uses all of the hydro that is available before using its steam plants. At times of high water a company may be able to generate more electricity from hydro power than it can sell to its customers. If the water is not run through the generating turbines it must be wasted over the spillway.

To prevent waste, the company will use the water to generate what is called *dump energy.* The company may sell this dump energy to another company that will buy it whenever it is available. Obviously the price of this dump energy is quite low—the lowest of all types of energy. Almost any price results in a gain to the seller. The price must be lower than the production cost of any of the next available stations of the buyer, or else the buyer will not benefit and will not buy. The rates are worked out to benefit both parties in order not to waste the energy.

5. *Emergency Power Contracts.* Neighboring companies usually interconnect their transmission systems, even though normally they do not need to buy or sell power. In an emergency one company can deliver energy to the other over this interconnection, and will supply whatever power it can to its neighbor after meeting the needs of its own customers.

6. *Stand-by Contracts.* A company may not have enough reserve capability and may want to buy a block of power from a neighbor simply as stand-by in case of an emergency. One company simply agrees to stand by for a certain number of kilowatts of capability for its neighbor at an agreed price.

Recent Evolution

By the beginning of World War II, a number of company groupings had already developed, each operating in *synchronism.* A group is operating in synchronism when all the generating units on these systems are running at a precise speed of some multiple of sixty as they are producing 60-cycle electricity. They may be operating at 1,800 or 3,600 revolutions per minute, for example. When operating in synchronism, all these units move together like a series of pendulums swinging in rhythm and in step, at 60 cycles per second. They all move together to the left and to the right with no one being out of step or out of phase. If one machine gets out of step, it is out of synchronism, and relays remove this unit from the system. Operating an interconnected system is a delicate process. For example, if a generating unit is to be connected to the system, it must be brought up to speed so that the frequency of the electricity it produces is the same as that of the system, and then it must be connected to the system in such a manner that it will be in phase or synchronism with the system. Once the machines are operating together they tend to be held in step by magnetic force.

During World War II electric utility companies were unable to purchase new generators as practically all the manufacturing capacity was being used to make turbine generators for ships and for other war purposes. Consequently, power companies had to operate all their plants, both the efficient and inefficient ones. They were able to carry the war load and the civilian load because they had developed interconnections. Without these, the industry could not have met the demand. In fact, at the beginning of the war there were many in the country who predicted that it would be impossible to meet all civilian and war needs for electric energy with the existing capabilities. Practically all commodities were being rationed. It seemed reasonable to suppose electricity would need to be rationed, too. Charles Kellogg, then President of the Edison Electric Institute, joined the Federal government and studied the situation. He estimated that the growth in use of energy during the war would not vary greatly from the trend then being experienced. This especially applied to the kilowatts of demand. He pointed out that load factors would be higher with factories employing a three-shift production schedule, but energy would be used in proportion to the material and manpower available for production. This meant, for example, that machines and men would be used to make tanks and trucks instead of cars. There would be a shift in the use of energy along with increased use rather than a tremendous increase in demand. This proved to be the case.

Certain situations did need special solutions. For example, it was decided to locate a large war industry in Arkansas. No one power company in the region had sufficient capability available to serve this new industry, nor could new capability be added. As a result, some twelve companies in the Southwest pooled their resources and together had sufficient capability to serve the industry throughout the war period. This was the beginning of what is called the Southwest Power Pool. It has been expanded in scope and size over the years and is still in existence. Other such pools were being formed in other parts of the country.

Throughout the war period the electric power industry was able to meet war demands and civilian demands as well. J. A. Krug, Director of the Office of War Utilities, complimented the industry at the close of the war. "Power has never been 'too little or too late,'" he said. "I do not know of a single instance in which the operation of a war plant has been delayed by lack of electric power supply."

Power Pools. The term *power pool* generally refers to the pooling of one power supply system with that of a neighbor or a number of neighbors. Pools may be formed for any number of reasons, from a desire for information exchanges or a desire to share ownership of a large power plant to a need for doing all of the things listed above under the benefits of interconnection. To obtain the maximum economies, the aggregate load of a pool should be large enough to allow the use of very large generating units. This means that with a unit rating of 1,000,000 kilowatts, a high rating at the present stage of technology, the size of the pool should be in the neighborhood of 8–10,000,000 kilowatts or more, so that if such a large unit should fail or be shut down for overhaul the percent of the total capability down will be in line with what is considered normal reserve requirements. That is, if the size of the largest generating unit in the pool were very large in relation to the total load of the pool, an inordinate amount of reserve would be required and this added expense would defeat the purpose of obtaining the economies of scale inherent in the large unit.

Power pools may take a number of forms, depending on the benefits sought and the resources available.

1. A separate generating company may be formed which is owned by all companies in the pool. The companies then di-

vide the output of the generating company. The generating company is operated for the maximum benefit to all, and the companies share the savings.

2. Two or more companies may build a generating station jointly. In this case each company may own a part of the generating station, or one company may finance the station and own it and the other companies will contract to buy a portion of the output of the plant over a period of years.

3. A number of companies may agree to pool all of their generating plants and major transmission systems. Holding companies fit into this category, but the companies may be owned independently and agree by contract to build their systems as though they had common ownership and management and to operate them as one with the benefits equitably divided among them. A committee of executives forms the management committee, and a committee of engineers plans the requirements of the combined system. In this planning the idea is to build all future plants and transmission lines just as they would be built if one company owned all the facilities.

Each company may agree to pay for the construction in its territory, whether it is a power plant or a section of transmission lines, or other arrangements may be worked out for financing construction of transmission lines which are considered common to all. Each company agrees to operate and maintain the facilities in its territory even though they have been built for the benefit of the whole. A central dispatching office is established and a central dispatcher operates all power plants in the pool to realize the maximum economy for the whole. A system of payments is worked out so that no company pays more than its share of the costs involved and each company receives its share of the savings of the enterprise. Savings can be calculated on the basis of what it would have cost each system to build and operate its own plants individually as compared to the scheme of building the plants and lines and operating them as an integrated system.

Economic Loading. Power companies now use a method of economic loading of power plants. For any large interconnected system, whether it is one company or a pool, there are a number of factors affecting efficiency.

First, there are many generating units of varying efficiencies. The efficiency of each changes as the load on it is increased or decreased. Fuel contracts vary. Then there is the factor of distance from the power plant to the load center—the place where the electricity is to be used. Power plant A may have a lower production cost than power plant B, but if power plant B is nearer the load center, the greater transmission loss of A over B may offset this difference. Assuming a given load curve, engineers can calculate the most economical loading on each generating unit for each hour of the day in the year. However, this requires an enormous number of calculations. Complicated mathematical equations are involved. It may take months to make the calculations; nevertheless, the possible savings justify them.

Today computers make the calculations for the most economical loading on each of the generators on a continuing basis. Before the advent of computers the central dispatcher would call upon the various plant operators to meet the schedule called for by an over-all schedule of operation. In the modern power pool the computer assists the operation of the system. All data are fed into the machine, which then schedules the operation of all the power plants and the load they are to carry each hour. There may be scores of power plants covering a number of states under the central control of one computer. The central dispatcher has his own schedule to follow, and if anything should happen to the computer he can take over.

This machine calculates, automatically and almost instantaneously, which units should be operated and the load to be put on each unit in order to get the most economical operation for the whole system.

Chart 11.5 shows how the various generating plants may

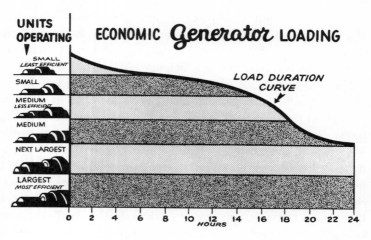

CHART 11.5

be called into service during the day in the operation of a large power pool. It also shows the loading on a load duration basis.

Pooling brings about the most economical construction and operation for the benefit of the companies in it. Savings are divided among all participants.

Studies to Form a Pool. In theory, before any pool is formed elaborate studies are made to ensure that pooling will result in good reliability, an over-all saving, and a proper division of the equities. In practice, a pool usually is formed as an extension of interconnection agreements that already exist among the potential members of the pool. However, to understand the workings and economics of a pool it is necessary to discuss the economic studies in some detail.

First, studies are prepared as to the pool size. A pool might be so large as to be unmanageable from a practical standpoint. It also can be so small as to be impractical. Computer studies help management determine the proper size and the companies which should be included. The load characteristics of neighboring systems have some bearing on the benefits of pooling, and all the systems want to obtain the maximum benefits including any diversity.

After a determination is made as to which companies will form the pool, studies are prepared to indicate the possible savings. This requires a ten- or twenty-year forecast of the kilowatt-hours and kilowatts for each of the companies separately. Each company is then analyzed to determine how much it would cost to build and to operate an expanded system of lines and plants to serve each of the companies separately and to operate them as separate units. To make the study a design must be prepared for each of the companies for a ten- or twenty-year period. A calculation must be made on economic loading of each of the systems separately for a period of ten years or twenty years with the economic loading by hours. Without the computers such a study would be too long to be practical.

Then various plans, A, B, C, or D may be prepared for the building of the plants and lines to serve all companies as a whole for a period of, say, ten years or twenty years. Then economic loading is applied by hours under each of the plans for the period under study. By this process the most economical of all the over-all plans can be determined. Thus a calculation can be made as to the savings that can be derived under any of the plans studied as compared to building plants to serve each company separately.

If the managements agree that the savings are justifiable they enter into negotiations of how these savings are to be divided equitably among them. These contracts have to be approved by appropriate regulatory bodies. They must be fair to all, and they must avoid any unjust discrimination.

There are various ways that these pools can be formed, and there are various ways to work out the equitable division of the savings.

Local conditions determine how best to work out these arrangements.

Pools to Serve Large Industries. At times electric utility companies are called upon to serve customers with unusually large loads, such as the Atomic Energy Commission (AEC). Two principal cases of this kind are discussed below.

At Joppa, Illinois, five power companies combined their resources to finance the installation of a 960,000-kilowatt plant for the Atomic Energy Commission. The total plant cost over $182 million. In Ohio and Indiana a similar project has been financed by fifteen power companies serving parts of eight states of the greater Ohio Valley Basin. The Ohio Valley project has installed 2,365,000 kilowatts of capability. It was also built to provide electricity for the Atomic Energy Commission. The cost was over $356 million.

In these two projects, power companies have about half a billion dollars invested to serve a single customer. The contracts, however, provide for cancellation charges should this customer cease operations. These charges are designed to permit the participating companies to absorb this capability on an economical basis.

Period Following World War II

The period following World War II presented another set of circumstances and further opportunities for innovation and development of the large interconnected systems. At the beginning of this period the power companies had very low reserves. They had been unable to purchase turbine generators during the war. Because of the higher cost of operating older plants and because the newer plants had not yet had their full effect, a number of companies were required to apply to their respective commissions for increased rates, which helped keep earnings at a satisfactory level so that the additional plants on a large scale could be financed in the market. As the charts in Chapter 5 show, this period witnessed a large increase in the required investment to serve the ever-increasing demands for electricity. Unlike World War I, when loads decreased with the cessation of hostilities, loads continued to grow.

This period was also characterized by a substantial increase in the size of generators purchased. Frequently the savings realized by these new large units helped finance new plants. Thus, the companies were encouraged to reach out to still larger units in order to realize still further economies which could be reflected either in a slowing of rate increases or in actual rate decreases.

With the movement to large generating units companies were required to extend further their transmission lines into higher and higher voltages and to build more and more transmission lines with more interconnections among systems.

This expansion into very large units and extra-high voltages presented new problems, just as every step along the evolutionary process presented problems.

Beginning in the late 1950s and early 1960s the evolutionary process was resulting in the formation of a number of large interconnected and coordinated areas throughout the country. The companies in the Pennsylvania, New Jersey, and Maryland area, known as the PJM Interconnection, had been working together for years. The large holding companies were coordinated among themselves and were planning coordination with neighboring groups. The Northwest Power Pool had been in operation for some years. Here the investor-owned companies and the non-investor-owned systems, including the Bonneville Power Administration, interconnected and coordinated the operation of their systems for maximum reliability and economy. The California companies were interconnecting and pooling. The Southwest Power Pool had been in existence for a number of years. The Texas companies were coordinated. Other areas were moving along the same line.

At about this time, the power companies formed a Committee on Power Capacity and Pooling in their national trade association, the Edison Electric Institute. A task force was appointed to study the interconnected and coordinated arrangements being made throughout the country so that all companies and all areas could benefit by the experience of other companies and other areas. A series of meetings was held around the country where these interconnection principles were discussed and the problems reviewed and studied.

Coordination of Power Supply. When companies interconnect their systems those systems operate in synchronism. Electricity flows with the speed of light. Any mishap on one system can potentially and instantly affect neighboring systems. The systems are designed to withstand almost any mishap. That is, if lightning strikes one line, the system is so set up that power automatically is shifted to another line without the customer knowing it. Generating units fail from time to time, but the system is designed so that such units are automatically cut out of service and power is supplied from other plants. On rare occasions there are even human failures, where an operator may fail to open or close a switch, which might cause a line failure or plant failure. But here again automatic devices isolate the defect and reroute the energy so that there is no interruption to the customers.

Before systems are expanded, the designs are placed on a computer so the engineers can see how they will work. This, in effect, is like building a model system. Engineers study this model and make sure that if reasonably expected failure occurs in one area, the automatic relays cut that portion of the system out and the remaining facilities meet the load. They study current performance as well as performance planned for the years to come.

Chart 11.6 shows how coordinating areas were developing during the early 1960s. They were evolving out of existing interconnection and pooling arrangements which had been developed over the industry's long history of interconnection and cooperation among systems. This kind of coordination also requires that all members of the coordinated area be fully informed as to what is being done by every other supplier in the system.

COORDINATION AREAS / 1960's

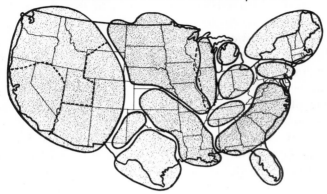

CHART 11.6

Northeast Power Interruption

On November 9, 1965 the power suppliers of the Northeast were operating their systems on an interconnected basis. There were ample reserves and there was no power shortage. But instability was triggered by the sudden loss of five transmission lines carrying power from the Sir Adam Beck Power Plant of the Hydro Electric Commission of Ontario to the Toronto, Canada, area. The power delivered over those lines was above the backup relay settings, thus causing the circuit breakers to open. With the loss of these lines, 1,500,000 kilowatts of power being generated at the Beck plant and at the Power Authority of the State of New York's Niagara plant, which had been serving Canada's loads, reversed and was superimposed on the lines south and east of Niagara. This quickly initiated periods of overgeneration, undergeneration, and instability which resulted in the shutdown of the systems in the Northeast.

Fundamentally, this is what happened. Prior to the mishap all systems were operating in synchronism; that is, all generators were running precisely at multiples of the same speed and all "pendulums" were in step swinging 60 cycles per second. As has been noted, when in step all machines tend to hold each other in step until a disturbance or a change in generation requirements throws them out of step.

The manner in which electric power systems behave when the energy balance between generation and load is changed is complicated by the fact that not only are there many generating plants and transmission lines in a system, but there also are many systems in an interconnected network. Fundamentally the reaction of the interconnected systems and their facilities depends on the relative strength of the various transmission lines, the size of each generator, each generator's control equipment, and the tie-line control between the various interconnected systems.

A full explanation of these factors is beyond the scope of this book. However, the principle can be understood by considering a system with a single plant and considering only the reaction of the generating plant equipment.

For example, as customers increase their load, the power demand on the power plant increases. Increased load will slow down the generators with a consequent decrease in their frequency. This is not appreciable with small load changes. Automatic devices called *governors* are installed on generating units to call for increased fuel so the plants can carry increased load and so the generators will not slow down appreciably and not get out of synchronism. There is a time lag between the increased load on a plant and the time the governors can react by adding fuel and steam to hold the generators at synchronous speed. If the load change is not too great, normal operations will swiftly bring the machine back to synchronous speed and there will be no appreciable or even recognizable effect on frequency. Under these conditions the system is said to be *stable*.

If a large load, such as 1 million kilowatts, is suddenly dropped from the plant there is not time for the various devices in the plant to react. The energy balance is destroyed and the extra fuel and steam cause the generators to speed up. As it speeds up, the frequency increases. In actual systems the energy balance mismatch will differ from unit to unit depending on the factors previously mentioned. Power generating units are equipped with automatic protective devices which can cut the units out when their frequency goes above or below a predetermined level.

When, on November 9, 1965, the generating units serving the load in the Toronto, Canada, area were disconnected from this load, they accelerated with a sharp drop in their electrical output. But as the speed increased their electrical power output increased rapidly. However, it was out of step with the other generation and this resulted in an unstable situation.

In this way the whole Northeast experienced an interruption of power supply in a matter of minutes. Each operator then examined his power supply and transmission equipment to determine the fault. There was no sign of equipment break or failure. Each system then began the complicated process of restarting its generators and restoring service. The restoration took longer in some areas than in others because of the intricacies of the systems.

The Northeast interruption called for a reevaluation of all coordination activities in the bulk power supply. Consequently, the best experts in the industry, acting through the committee structure of the Edison Electric Institute, made a complete study of industry coordination practices. They prepared a report on coordination of electric power supply which was made available to all power suppliers. It contained the principles of coordination given in the appendix of this book. Of course, every power supplier and coordinating area made its own reevaluation of its local power systems.

Coordination and Pooling

Sometimes the terms *coordination* and *pooling* have been used interchangeably. However, there is a growing tendency to attach special meanings to each of them. For clarity, the terms will be defined as they are used in this book. Generally speaking, this is the manner in which the terms are now used in the industry.

Area coordination is concerned only with the reliability of bulk power supply and not with economies available through pooling and other arrangements. Pooling has to do primarily with the economies of planning and operation.

Coordination areas are usually larger than pooling areas. Coordination involves a larger number of power suppliers. In fact, there may be a number of pooling areas within a single coordination area. A pooling arrangement usually involves a smaller number of suppliers. As pooling contracts among suppliers involve sharing economies of scale and possibly joint building of facilities, the negotiations leading to contracts are far more complicated and time consuming than for coordination contracts. Further, regulatory bodies need to examine the pooling contracts thoroughly to be assured that all parties equitably share in the benefits of pooling.

The coordination area groups have evolved from existing organizations and differ as to size of area and number of systems because of differences between areas as to total population; population density; the magnitude, nature, and density of electrical power loads and resources; extent of interconnection; and other factors.

Chart 11.7 shows the coordination areas as they existed in 1968.

Table 11.1 shows the number of investor-owned and non-investor-owned systems in each of the areas, the size of each area in square miles, and the aggregate generating capability of each. Each coordination area is designed to do those things most fitting to meet local conditions and to provide the most reliable service. Generally speaking, the participants in each coordination area include all power suppliers that make substantial contributions to the delivery of electric energy in the area, whether the system is financed in the market or through the government. These suppliers agree among them-

COORDINATION AREAS / 1968

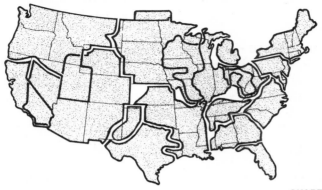

CHART 11.7

selves on the procedures, principles, and practices to be followed.   Usually there is a signed agreement among the parties.   Each power system in the area agrees to do all those
things on its own system to be sure that it gives good service
locally and does not operate in a way that can affect the
other systems detrimentally from the standpoint of reliability.
Coordination does not mean that an individual utility no
longer plans or builds its facilities to serve its own customers.
Coordination supplements this process, giving each local system more scope for improvement.

The various coordination areas are in different stages of development, but basically they are all considering the establishment of planning and operating criteria that can be used
in simulated testing and evaluation of the performance of the
combined systems.   Each member discloses to the other
members his plans for major generating and transmission facilities in advance of construction, so that the effect of these
new facilities on the reliability of the area may be evaluated
and where necessary, modifications made to assure the reliability of the bulk power supplier of the entire area.

The coordination areas also are developing procedures for
the systematic interchange and review by all the participants

of information pertaining to load projections, significant changes in system facilities, modes of operation, and all other matters pertinent to the reliable performance of the area systems.

TABLE 11.1   Statistics on Area Coordinating Groups

| Coordinating Group | Number of Systems | | | Year-End 1968 Capability (Million Kw) | Size of Area (Sq Mi) |
	Investor Owned	Non-Investor Owned	Total		
Northeast Power Co-ordinating Council	17	3	20 [1]	40.00	271,000 [2]
Mid Atlantic Area Coordinating Group	12	0	12	23.34	48,000
Carolinas-Virginias Power Pool (CARVA)	4	0	4	13.91	94,000
Southern System	4	0	4	10.16 [3]	122,000
Florida Group	3	2	5	9.30	45,000
East Central Area Reliability Group (ECAR)	25	1	26	43.11	192,000
Southwest Power Pool	20	3	23 [4]	21.65	300,000
Mid-America Interpool Network (MAIN)	18	1	19	28.30 [5]	280,000
Mid-Continent Area Reliability Coordination Group (MARCA)	15	11	26 [1]	14.42	350,000 [6]
Texas Interconnected System	6	3	9	16.19	195,000
Western Systems Coordinating Council (WSCC)	19	21	40 [1]	60.50	1,430,000 [7]
Tennessee Valley Authority (TVA)	0	1	1	18.20 [8]	80,000

NOTES:

1. Includes as a member a Canadian electric power system or systems.

2. Includes the service area of the Canadian system which amounts to 155,000 sq. mi.

3. System peak load.

4. Participants and contributing members.

5. Does not include the liaison members.

6. Includes southern portion of Manitoba Hydro service area.

7. Includes the service area of the Canadian system which amounts to 330,000 sq. mi.

8. Generating capacity.

Reserve capability policies are given careful consideration in each coordination area.   Each coordination area considers maintenance scheduling of major generation and transmission facilities so that such maintenance does not jeopardize service reliability.   Coordination areas also agree upon essential metering, communication, and relaying facilities.

Some coordination areas have established a coordination office and maintain a staff devoted solely to coordination of bulk power supply in the area.   Through this office all information and all studies are coordinated so that all operators and planners of the coordinated systems can always be kept informed.

Most area groups have an executive or management committee to review principles and procedures on matters affecting reliability.   This committee also has the responsibility to make certain the planning for generation, transmission, and other relevant matters of each of the parties is reviewed and evaluated.   Assisting this top committee are one or more technical committees.   In most area organizations, task force or ad hoc committees of specialists assist the standing technical committee or committees.   These technical committees generally perform studies and investigations concerning such things as over-all adequacy of transmission facilities, generation reserves, operating practices, and procedures.

Coordination Between Areas of Coordination.   Interconnection between power systems is not limited to those within each coordination area.   Power systems in adjacent coordination areas may have interchange contracts with each other.   Inter-area coordination agreements supplement bilateral and multilateral interconnection arrangements.   It is of interest to each coordination area to know the principal facts about adjoining areas.   Consequently, procedures are established as seem appropriate to local conditions to bring about the proper exchange of all knowledge and information between areas.   Also, through the industry's trade association there is a continuing exchange of information and ideas so that each area and each system can know of the practices, experience, and procedures of other areas.

For example, a committee of the Edison Electric Institute maintains a continuing file on the contractual arrangements for the purchase and interchange of power between power systems so that all members can review them. Through the Institute, exhaustive studies of diversity are made among all power suppliers. These studies cover a period of years and take into account the hourly demand on individual power systems and their neighbors. As an example of the benefits of diversity and how they can be realized, there is a contract between a group of power systems in the Southwest Power Pool called the South Central Electric Companies (SCEC) and the TVA. SCEC has a heavy peak in the summer due to air conditioning. TVA has a heavy peak in the winter due to electric house heating. There is a significant annual load diversity between them. A contract calls for delivery of capability by the Southwest companies to TVA during the winter and for the delivery of an equal amount of capability by TVA to the Southwest companies during the summer. Both suppliers and their customers benefit.

Energy Sources

Power companies are interested in making electricity at the lowest possible cost. They use hydroelectric power whenever it is economically available prior to using steam-generated power. Chart 11.8 shows the percentage of hydro and steam generation by years for the investor-owned electric utility industry since 1930. The percentage of hydro will likely continue downward.

Hydroelectric Power. Hydroelectric power is made from falling water which is used to turn turbines connected to generators. The higher the fall, the more electricity can be generated from a given volume of water. If the flow of water is continuous, the hydroelectric station can make energy day

ELECTRIC UTILITY COMPANY GENERATION

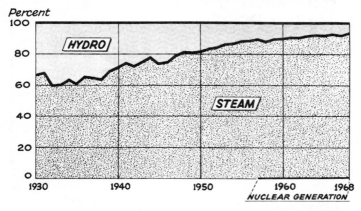

CHART 11.8

and night, month in and month out.   This will be possible
where a dam is built on a river with a constant stream flow
or where the reservoir behind the dam is large enough to
store up a reserve of water to run the turbines continuously.

The number of dam sites of this kind is limited.   Since
these are the best kind, they were the first to be developed.
After these had been developed, engineers began to work on
the next best sites for dams.   Most of the practical dam sites
have by this time been developed.   The remaining dam sites
are largely in places where the stream flow is limited and er-
ratic.   There may be an abundance of water during the rainy
season, but there may be little, if any, during the dry season.
At sites such as this, the water is stored in an artificial lake
created by the dam, where storage is possible.

When a dam is built across a river with limited stream
flow, there are two principal ways that the power can be
used.   On rivers such as these there is sometimes a great deal
of water—even floods—and at other times the rivers are almost dry.   These rivers can make a large quantity of electricity for short periods and much less electricity during the rest
of the year, depending upon how the water flows and the size

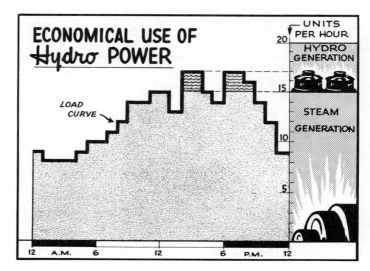

CHART 11.9

of the reservoir behind the dam.   If the company tried to use the generators at these dams to make electricity twenty-four hours a day all year long, a good portion of the capacity of the plant would be standing idle, because the water would have to be stretched out over the whole year.

Sometimes the water is stored up and the generators are operated only a few hours a day and a few days out of the year.   Where this is done, much more electricity can be produced for every hour in which the generators are operating.[2]

The economical way to use this kind of hydroelectric power is to save up the water during the period when people are not using much electricity, and then run the generators at their full capacity when people are using a great deal of electricity during the peak hours of the day.   This is illustrated in Chart 11.9.   At times of lower demand, steam stations take care of the load, and a supply of water is built up.   Then

---

[2] Many dams are built by the government for flood control, navigation, irrigation, and power.   See *Government in the Power Business* by Edwin Vennard, McGraw-Hill Book Company, 1968.

when the peak hour arrives, the water is released through the turbines and the power is poured into the system.

Combination Dams. At times a dam is built to serve a number of different purposes. It may be a dam for both flood control and power generation or a dam for irrigation and power generation. Navigation improvement may be one of the functions. At times the various functions operate at cross purposes. Then it is necessary to establish the primary purpose and secondary purpose.

Chart 11.10 shows a flood-control dam. It needs an empty reservoir to catch the floods. Following each flood, the reservoir empties through openings near the bottom at a rate not greater than the river channel can take without overflowing. But, of course, no power can be obtained from an empty reservoir, and that is the normal condition in a dam built strictly for flood control—the reservoir is kept empty.

Chart 11.11 shows a power dam. To get power, the water must fall; the greater the fall, the greater the power. Therefore, at a power dam the reservoir is kept as full as practical.

CHART 11.10

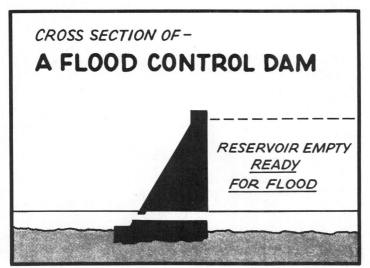

CROSS SECTION OF –
A FLOOD CONTROL DAM

RESERVOIR EMPTY
READY
FOR FLOOD

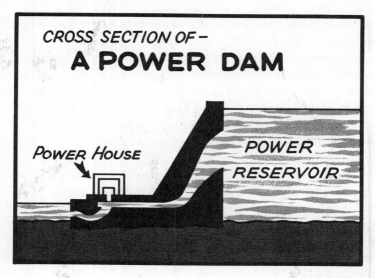

CROSS SECTION OF –
A POWER DAM

Power House

POWER
RESERVOIR

CHART 11.11

CHART 11.12

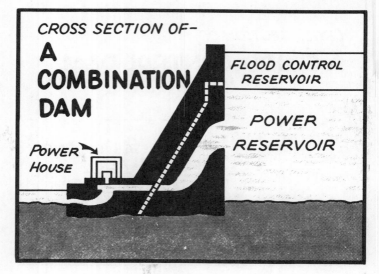

CROSS SECTION OF–
A
COMBINATION
DAM

Power House

FLOOD CONTROL
RESERVOIR

POWER
RESERVOIR

Unlike a flood-control dam, there are no flood discharge openings near the bottom.

Chart 11.12 shows a combination dam built to provide both flood control and power. This really means two projects at the same place—one for power and one for flood control. A dam of this kind has to be considerably larger and higher than either a flood-control dam or a power dam separately. Likewise, it is more costly.

Cost of Hydro versus Steam. Engineers know the relative costs of hydroelectric power as compared with steam power, but many people who are not engineers believe that electricity made from falling water is always cheaper than electricity made by burning fuel. At first glance it may seem that hydroelectric power should always be cheaper, because one has to buy fuel for a steam plant, whereas the water is free. However, this overlooks the fact that variable costs, such as fuel cost, are not nearly as important as fixed costs. The fixed costs of a hydro plant are likely to be much higher than for a steam plant.

In the early years of the power business hydroelectric power was generally cheaper than steam power. However, designers and manufacturers have been able to raise the efficiency of steam generation and hold to a minimum the increase in the unit cost of the machines. There has been less opportunity for raising the efficiency of hydro plants. As a result there has been a shift in the relative economy of the two over the years.

Following are the principal handicaps of hydroelectric power:

1. In most cases the hydroelectric plant requires a much higher investment per kilowatt of capacity. In both steam plants and hydro plants there must be a turbine and a generator. But the hydro plant calls for an expensive dam, and the company usually has to buy considerable land for the reservoir.

2. The hydroelectric plant must be built at a suitable site

on the river. The site may be some distance from the point where the power is needed. This calls for a higher investment in transmission facilities. Usually the steam plant can be located nearer the load centers.

3. Frequently the hydro plant depends on stream flow, which is sometimes intermittent. In many cases the hydroelectric plant cannot be used around the clock every day in the year. It must be supplemented by steam power.

4. The steam plant offers greater opportunities for improvement in design. Witness the fact that in the early days of the industry it took about 8 pounds of coal to make a kilowatt-hour in a steam plant. This compares with less than two-thirds of a pound per kilowatt-hour in a modern steam plant. Also, the investment cost per kilowatt in the steam plant has been fairly steady, as noted in Chapter 5.

5. Frequently the hydro dam is built to carry out some other purpose, such as flood control, in which case the two functions may work at cross purposes, as noted on page 281.

As far back as 1940 this trend in favor of steam plant generation was noted by Mr. Gano Dunn, president of J. G. White Engineering Corporation, when he was serving as power consultant to the Advisory Commission of the Council of National Defense. He summed up his viewpoint as follows (Report No. 1953, accompanying Senate Joint Resolution 285):

> And today, in 1940, a new plant can produce a kilowatt-hour for 0.9 pounds of coal. In that scaling down from eight pounds of coal per kilowatt-hour to 0.9 pounds of coal per kilowatt-hour, steam power passed water power, and is now much the cheaper power when costs are calculated on the same basis for each form of power.

Table 11.2 shows a comparison of steam and hydro and how the change has come about in the past 48 years.

Pumped Storage. The pumped storage process offers the opportunity for economic development of power in areas where the terrain is suitable. Pumped storage requires that there be an elevated location where a reservoir can be con-

TABLE 11.2   Comparison of Steam and Hydro*

		1920	*1968*
a.	Hydro:		
	Investment in plant		
	(including 20% for transmission)		
	per kilowatt	$240	$265
	Fixed charges:		
	Return on investment	8.0%	8.0%
	Depreciation	1.5	1.9
	Taxes	1.4	1.9
	Total	10.9%	11.8%
b.	Steam:		
	Investment in plant per kilowatt	$ 140	$ 110
	Fixed charges:		
	Return on investment	8.0%	8.0%
	Depreciation	2.5	2.8
	Taxes	2.0	2.7
	Total	12.5%	13.5%
	Economy, Btu per kilowatt-hour	30,000	8,700
	Coal:		
	Pounds coal (of 13,000 Btu per lb)	2.31	0.67
	Cost per ton, dollars	4.50	7.25
c.	Load characteristics (55% capacity		
	factor), kilowatt-hours per kilowatt		
	per year	4,820	4,820

Costs, Mills per Kilowatt-hour

		1920	*1968*
a.	Hydro:		
	Fixed charges	5.43	6.49
	Fuel	0.00	0.00
	Labor and maintenance		
	(including transmission)	0.35	0.89
	Total	5.78	7.38
b.	Steam:		
	Fixed charges	3.63	3.08
	Fuel	5.20	2.43
	Labor and maintenance	1.40	0.85
	Total	10.23	6.36

 * This table shows that a kilowatt-hour of hydro-generated electricity costs 7.38 mills, while for steam power the cost is only 6.36 mills.

structed. There needs to be a water supply such as a river or lake nearby. A hydroelectric power plant is built at this site. It might be as large as 1,000,000 kilowatts. During the off-peak hours on the power system the hydroelectric plant is run as a motor with the energy fed to it by the power system supplied by base load steam plants. The motors run the water wheels and pump the water from the river or lake up into the reservoir until the reservoir is filled. During the peak hours the water in the reservoir is released and is run through the water wheels of the turbines. The plant operates as a generator of power. This method is economical because, as a rule, the cost of building the pumped storage plant and reservoir is less than the cost of an ordinary hydroelectric station and *less* than the usual cost of a steam plant. Engineers believe the combination will be especially economical when power systems have nuclear steam power which will have unusually low energy cost. Thus the low energy cost can be used to lift the water into the reservoir, and the pumped storage plant can operate at full capacity during the period of peak demands, thereby reducing the over-all investment required to meet the peak.

Nuclear Power. Until recently the principal raw energy sources have been the fossil fuels, namely coal, gas, and oil. Although America has an abundance of these fuels, reserves are not inexhaustible. With the rapid rate of increase in use of energy on the part of the American people it has been estimated that the economic sources of oil and gas may be exhausted sometime within the next century. The supply of coal will last much longer.

Fortunately, during the late 1940s a new source of raw energy, nuclear energy, appeared and it has enormous possibilities.

Up until 1954 the principal knowledge concerning nuclear energy was in the hands of government. In 1954 an act of Congress made this knowledge available so that businessmen and others could work toward controlling the energy so as to make it useful for peaceful purposes.

Then followed a massive cooperative research and development effort of the electric power industry, its equipment manufacturers and suppliers, consulting organizations, and the United States Atomic Energy Commission. As a member of the Edison Electric Institute, the Edison Power Company cooperated in this research and development. In 1958, the Institute appointed a Task Force on Nuclear Power composed of several of the nation's leading scientists in the field as well as knowledgeable members of the power industry. Its task was to study the various methods that might be utilized to convert this new source of raw energy into thermal energy that could be competitive in the production of electricity with thermal energy produced with other forms of fuel. It recommended a number of possibilities. No one knew whether any of the various reactor types then being considered would be economically feasible. It was necessary to do further research on the most promising concepts and to build first reactor experiments and small prototypes.

The Atomic Energy Commission had set as its primary objective the attainment of competitive nuclear power in high-cost fuel areas by 1968. This would be fourteen years after industry was given access to nuclear technology with the passage of the Atomic Energy Act of 1954.

There followed elaborate research and development efforts on the part of many organizations with a commitment of roughly $2.5 billion for the development of this new source of fuel. Through 1968 the investor-owned electric power industry spent or committed almost $1 billion to nuclear research and development with Edison Power Company paying its part.

In 1964, four years ahead of when the AEC hoped that nuclear power would be competitive with fossil fuels in high-cost fuel areas, a nuclear power plant was purchased on the basis that it would prove to be competitive with a conventional plant of the same general size and at the same location.

By providing the Edison Power Company with another source of raw energy, nuclear fuel stimulates competition with and among the fossil fuel industries—all to the benefit of the electric utility customers. Even though nuclear power accounts for only a small fraction of the electric energy now produced in the United States, its competitive effect already has been felt. For example, complete fission (the splitting of the atom) of one pound of uranium would liberate roughly the equivalent energy produced in the combustion of 2,400 tons of coal. The cost of delivering nuclear fuel is insignificant and, as a consequence, the distance from the source of the nuclear fuel supply is not an important factor in the cost of producing electric power by this means. This means that those areas of the country with high freight costs, because of their distance from raw fuel supplies, may now especially benefit from this new nuclear fuel.

How a Nuclear Plant Works. A nuclear power plant is in many respects similar to a conventional fossil-fuel-burning plant. The chief difference is in the way heat is generated, controlled, and used to produce steam to turn the turbine generator.

In a nuclear power plant the furnace which is used for burning coal, oil, or gas in a conventional plant is replaced by a reactor which contains a core of nuclear fuel. Energy is produced in the reactor by a process called fission. In this process the center or nucleus of certain atoms, upon being struck by a subatomic particle called a neutron, splits into fragments called fission products which fly apart at great speed and generate heat as they collide with surrounding matter.

The splitting of the atomic nucleus into parts is accompanied by the emission of high-energy electromagnetic radiation and the release of additional neutrons. The released neutrons may in turn strike other fissionable nuclei in the nuclear fuel, causing further fissions.

A nuclear reactor is a device for starting and controlling a self-sustaining fission reaction. The nuclear core of the reactor generally consists of fuel elements in some chemical form of uranium and thorium or plutonium, depending on the type of reactor. Heat energy is produced by the fissioning of the nuclear fuel. A coolant is used to remove this heat energy from the reactor core so that it can be utilized in producing electricity. Fuel elements for water-cooled reactors are metal tubes containing small cylindrical pellets of uranium oxide.

Also under evaluation for commercial power production is the gas-cooled reactor, in which the fuel elements are fabricated basically of a uranium carbide compound and of a graphite, which acts as the structural material as well as the protective enclosure for the fuel material. The protective enclosure, whether graphite or metal, is called cladding.

To control the rate at which fission occurs, most reactors regulate the "population" of neutrons in the core. This is done mainly by rods which, when inserted into the core, absorb neutrons and retard the fission process. If the operator wishes to increase the power level or reaction rate, the regulating rods are withdrawn. To shut down the reactor, the rods are fully inserted.

Neutrons liberated in the fissioning process travel at very high speeds. This is not desirable in some reactor systems, where slow-moving neutrons are more effective in triggering fission than are high-velocity neutrons. To attain the desired slow-moving neutrons, a material called a moderator is used. It slows down the neutrons and has a minimum tendency to absorb them. Materials used for this purpose include graphite or ordinary water, the latter also serving as a coolant. Most power reactors presently in operation or under construction utilize the slow-moving neutrons and are called *thermal reactors.*

In addition to water, other materials, such as gas or molten sodium, act as coolants. The coolant transfers heat from the reactor to produce steam. In some types of reactor plants using water as a coolant, the water is allowed to boil in the reactor and the resultant steam is used directly in the turbine.

In the water-cooled reactor the nuclear fuel which has

been compacted into uniform pellets is placed into tubes or fuel elements. These fuel elements are then sealed at the top and bottom and arranged by spacer devices into bundles called *fuel assemblies.* The spacer devices separate the fuel elements so as to permit coolant to flow around all of the elements in order to remove the heat produced by the fissioning uranium atoms. Scores of fuel assemblies, precisely arranged, are required to make up the core of the reactor.

This geometric arrangement is necessary for several reasons. Nuclear fuel, unlike fossil fuel, has a very high energy density—that is, tremendous quantities of heat are produced by a small amount of fuel. Because of this the fuel must be arranged in such a fashion as to permit the coolant to carry away the heat. This requires that the fuel be dispersed rather than lumped together in a large mass.

Chemical reactions between the fuel and the coolant must be avoided and, as a safety precaution, the radioactive materials produced must be contained. For these reasons the fuel is contained in individual tubes or fuel elements. The cladding material from which the tubes are made must meet rigid specifications. It must have good heat-transfer characteristics, it must not react chemically with either the fuel or the coolant, and it must not unduly absorb the neutrons produced in the fissioning of the fuel to the detriment of continuing the chain reaction. The cladding material generally used for this purpose is thin-walled stainless steel or an alloy of the element zirconium.

The gas-cooled reactor, which utilizes an inert gas, helium, as the coolant, has a different core structure than the water-cooled reactor. The fuel elements are fabricated of graphite, which acts as the structural material and the neutron moderator as well as the cladding material. The nuclear fuel, consisting of both uranium and thorium, is formed into the center of the fuel element. Because the coolant is inert, the graphite serves as adequate cladding for the nuclear fuels. The inert gas will not react with or corrode the graphite or any other structural material. Physically, the fuel elements for the gas-cooled reactor are much larger than those of the water-cooled reactor and are not bundled into fuel assemblies but are individually arranged and spaced so as to allow coolant to flow around all of the elements, removing the heat produced by the nuclear reaction. Several hundred fuel elements are required to make up the core of the reactor.

The nuclear fuel, in the form of either fuel assemblies for the water-cooled reactor or fuel elements for the gas-cooled reactor, is placed in the reactor to produce heat which, in turn, is converted into electricity. After several years of operation a nuclear core must be replaced. By this time the absorption of neutrons by the accumulation of fission products is so great that there are too few neutrons remaining to maintain a chain reaction.

After the fuel assemblies or elements are removed from the reactor, they still contain material which must be reclaimed because of its economic value. In fact, only about 1 to 3 percent of the uranium has been "burned." The other 97 to 99 percent is locked in the hundreds of fuel elements of the "spent" core.

Breeder Reactors. Much remains to be done before the full benefits of nuclear power are obtained. One of the most promising types of nuclear reactors is the *breeder reactor*. This reactor gives further promise of economical energy because it breeds more nuclear fuel than it uses. The breeder reactor's fuel costs should be low and relatively independent of market fluctuations of uranium ore. According to some experts, without breeders the world might run out of low cost uranium by the year 2000.

While present commercial reactors use only 1 to 3 percent of the potential energy of the fuel, breeders offer usage of 80 percent or more of the potential energy. Edison Power Company, in cooperation with other power suppliers, manufacturers, and the Atomic Energy Commission, is going forward with research to develop the breeder reactor. The magnitude and cost of this research and development effort will be approximately as great as that for making nuclear energy commercially feasible. It is now contemplated that commercial breeder reactors can be producing energy for use by the customers during the middle 1980s.

Fusion. Fusion takes place when certain light atomic elements are joined or fused together with a release of heat energy. It is now known theoretically that the energy released by fusion can be controlled and released over a period of time. However, the problems in the research and development of controlled fusion are far greater than those experienced in the control of fission. Fusion takes place at a temperature of millions of degrees. To control this enormous energy, scientists must find some way to contain and control such temperatures.

12

The Company
and the Community

The business of providing electric energy touches the life of a community in many ways. The lives of the electric utility company and its customers are closely intertwined. The company must serve the needs of its area; it cannot move to another location. More than any other business, the utility company is dependent upon the goodwill of its customers and on the economic strength of the area it serves.

The Edison Power Company, like other utility companies, works to live up to its responsibilities as a community citizen. As the largest taxpayer in most of the communities it serves, it is a major contributor to public education and to the whole range of local governmental activities. In addition, like other businesses, it is called upon to make contributions to charities, community activities, and organizations. The company encourages its employees to take part in community affairs, knowing that whatever they do to improve the area in which they live will help the company prosper in the long run. In all matters, the company tries to find those things it does which its customers like and to do more of them while doing less of those things the customers do not like as well.

Economic Development

The company is concerned with the over-all economic development of the territory it serves. To help the area grow and to bring in more jobs, the company has established an Area Development Department. This department works to bring new industries into the area. It also works with local and regional planning bodies toward community improvement. For example, the company's Area Development Manager learned that a small precision parts manufacturer required a building immediately. After consulting with a regional planning group, the company saw that a particular community in its service area would benefit from the new employment. To ensure that the plant would be located in that community, the company turned over one of its own garages to the manufacturer for use until a permanent plant could be built. As a result, thirty new jobs were brought to the area. Of course, the company is equally concerned with the development of its rural areas and carries on educational services for farmers, helping them to increase crop yield and income.

Environmental Impact

One of the most important areas of concern to the management of the company is the impact its actions may have on the environment. As the amount of electricity people use increases and the physical equipment required to satisfy electrical needs grows in size and complexity, these problems have become more intense. Here, again, the company management must work closely with local and regional planning groups as well as with regulatory agencies to be sure that it is operating in the public interest.

Questions of plant siting, overhead versus underground lines, appearance of facilities, and similar matters have been of continuing concern to electric utility companies from the beginnings of the industry. The cleanliness, silence, convenience, unobtrusiveness, and flexibility of electric energy have been among its most attractive qualities. The industry has worked to make the most of these advantages.

Like most electric utilities, Edison Power Company makes a policy of placing electric distribution lines underground in the congested downtown areas of the big cities it serves. It also endeavors to place distribution lines underground in new residential developments where it is economically feasible and when customers want it. Recent technological advances have made this almost as economical as building lines overhead, though differences in terrain have not given all utility companies the same advantage.

In established residential areas, the company makes a policy of improving the appearance of overhead lines by raising voltage levels, which reduces the number of wires needed, and by other methods. Where possible, it makes a practice of concealing the wires behind trees or buildings. This effort is part of the company's regular expansion program.

From time to time the company has studied the cost of converting existing overhead distribution lines to underground, using the latest technological knowledge available. The most recent study indicates the cost would amount to almost eight times the company's present investment in distribution lines. This parallels a national study made in 1968 by a special task force of the President's Citizens Advisory Committee on Recreation and Natural Beauty, under the chairmanship of Laurance Rockefeller. In its report to the President, the task force estimated the investment in overhead distribution lines as $18 billion. To underground all of them, the task force said, would cost on the order of $150 billion. The task force also pointed out that undergrounding transmission lines of any length is not practical today. The Edison Power Company has joined with other electric utility companies in a massive research program aimed at finding economical solutions to the problems involved in underground transmission.

Appearance of Facilities. Several years ago the Edison Electric Institute employed Henry Dreyfuss, the internationally known industrial designer, to prepare a series of good-looking transmission towers. Along with other companies in the electric utility industry, Edison Power Company has started using Dreyfuss-designed poles in locations where appearance is of special importance. The company is also taking great care in placing the towers so they will be most pleasing and treats forested and wild areas with special attention so that they may be spared as much as possible when transmission lines are constructed. Transmission routes are chosen to cause the least impairment to natural beauty.

For years the Edison Power Company has called on professional architects in the design of its generating plants and substations, and has employed landscape architects in planning the land around its facilities and in developing recreational areas at its hydroelectric sites. These recreational areas are used by the public in large numbers.

Clean Air. The company, and the entire electric utility industry, has been concerned about emissions from steam generating plants from the earliest days. Improvements have been made step by step. Until about a generation ago, "smoke" was considered the major problem. Development of more efficient fuel-burning equipment and furnaces did much to relieve this difficulty. The next problem was "fly ash," the cinder discharge associated with both stoker and pulverized forms of coal firing. This problem began to receive special attention in the early 1930s, and special equipment was developed to remove fly ash from furnace chimneys. Efficiencies of over 90 percent are now the rule and in the newest plants, efficiencies are up to 99.5 percent. The Edison Power Company has joined with other companies in the industry in seeking ways to make use of the large amounts of fly ash it collects. Major attention in recent years has been given to controlling the sulfur oxides and other gaseous emissions from power plants.

The Edison Power Company continues to work with equipment manufacturers to develop boilers with higher combustion characteristics, and much progress has been made along this line. The more efficient the combustion process, the more energy is taken from the fuel used and the less waste there is to discharge into the atmosphere. There have also been significant accomplishments in designing stack heights and nozzles so that plumes will be discharged with great velocity and enter the atmosphere in ways which will not affect areas where people live. Often this necessitates building a much higher stack than would normally be required, which increases construction costs substantially. Some utility plants have stacks as high as 1,200 feet.

Of course, the process of electrification of industry has had a great effect on the quality of the air in industrial regions. While air pollution control was not the primary reason for substituting electric motors for oil, gasoline, and Diesel engines, the result has been to decrease air pollution. Converting raw energy forms to electricity is a more efficient way of using the energy than direct use in engines. Combustion is more complete and there is less waste discharged. Today, about 85 percent of the mechanical horsepower used in manufacturing in this country is in electric motors driven with purchased electric energy.

The same kind of trend is taking place in heating. During the past ten years, the number of electrically heated homes and apartments has been growing in all parts of the country, and this trend can be expected to continue. As the isolated heating plants in homes, apartment buildings, and office buildings are reduced, the amount of air pollution they create will be decreased.

Most air pollution in this country is a result of the relatively inefficient combustion process involved in the internal combustion engine and in the other major means of transportation. Edison Power Company is testing electric vehicles, with the hope that they will one day be competitive with passenger cars. This will not be possible until substantial improvements are made in present battery technology so that electric cars may travel farther on a single charge than is now possible. In the meantime, the company is encouraging the use of electric fork lift trucks and other vehicles in industry and for recreational use wherever possible.

Thermal Effect. Steam power plants make use of water, and as a result electric utility companies have done considerable work to increase understanding of the effects of the water they return to rivers, lakes, and harbors. The improvements in efficiency of fossil-fuel-burning power plants have decreased the proportion of waste heat per kilowatt-hour which power plants discharge into their condenser cooling water. Edison Power Company makes careful measurements of the discharges made into condenser cooling water, seeing to it that the temperature stays within acceptable government-defined limits. It employs biologists to study the ecology of the waters it affects, and is continually looking for alternative methods of cooling condensers. Since 1962 the company has also been contributing to an Edison Electric Institute research project, conducted by Johns Hopkins University, which is aimed at gaining a better understanding of thermal effects on ecology.

Public Relations Effort

The company's relations with the communities it serves are the responsibility of several departments. Each month, a special committee of the heads of these departments is convened under the chairmanship of the Vice President of Public Relations. The chief executive of the company regularly participates in these meetings. At each meeting, the committee examines and discusses the impact of the company's activities on the public. They may discuss tree trimming practices, telephone manners of those in the company who talk with customers, courtesy in letter writing, the type and tone of company advertising, the appearance of company employees, the condition of company equipment, the handling of collection problems, operating problems that affect the public, and many other subjects. They also may discuss means of informing customers and employees of the reasons for certain company actions and the facts about company policies and activities. The company recognizes that good public relations cuts across all the departments of the company and that the responsibility of maintaining good public relations must be assumed by everyone in the company, from the top down.

Epilogue: Looking to the Year 2000

As earlier chapters in this book have indicated, people in the electric power business are used to looking ahead. Companies must make their plans for five, ten, and even more years in advance. The industry is evolving along the basic patterns described in this book.

There has been a trend toward larger and more efficient generating units, toward higher and higher voltages, and toward more interconnection and pooling. Large-scale atomic plants and the introduction of new generating concepts, such as the breeder reactor, will bring further improvements in the efficiency of converting raw fuels to electricity. The amount of fuel needed to make a kilowatt-hour will continue to decrease. Increased use per customer, technological progress, increased efficiency and reliability, and knowledge that there are ample fuel supplies (assuming the continued availability of sound financing in the future) give people in the industry confidence to predict that the long-term average price of electricity will keep on going down despite inflation, and despite some necessary rate increases. Remember, most rate schedules provide an automatic decrease in the average price for increased use of electricity.

Cities of the Future. Efforts are now being made to visualize what cities may be like in the year 2000 so that electric utility companies will be prepared to serve their energy needs. While the year 2000 may seem far off, it is within the career span of one man. Those involved in forecasts agree on a number of significant points:

High speed electric railways will connect one city with another. Most travel within cities will be in small electric cars —and in some areas these cars will be able to move onto automatic roadways and be carried at very high speeds. They will move people as rapidly and easily on a horizontal plane as elevators now move people vertically.

In downtown centers, travel will be by electrically powered walkways and similar devices. The internal combustion engine probably will be used mainly for long-distance or intercity driving.

Downtown areas will be completely climate controlled. The air conditioned shopping malls springing up around the country today are forerunners of these cities of the future. While some planners dream of domes covering whole cities, others do not think this will be possible or even desirable. There seems to be agreement that special purpose domes will be commonplace. Stadiums, playing fields, and even backyards will be covered with clear roofing and kept at comfortable humidity and temperature levels.

Near the city will be farm factories, each one specializing in its own product and each one having its own optimum environment. Industry will be clean and heavily automated. The variety of goods available will be much greater than is known today. There will be a greater sense of space, though the number of people in the cities will be greatly increased.

Homes of the Future. The home of the future will be different, too. Here are what some of the manufacturers are talking about:

■ "Picture frame" television screens as large as a living room wall. They will show commercial programs, some of them three dimensional. They will be turned on and off with a wave of the hand and used as an intercom system within the house or between houses.

■ Electric waste disposal. There are a number of schemes for this, all much more sophisticated than the disposals familiar today. General Electric Company predicts that beams of searing light will vaporize all the refuse in a household.

■ Home computers that will keep the bank account and household budget up to date. Computers will link the family, through a system of computer utility companies, with central information banks.

■ Floor cleaning will be all-electric, probably with the use of automatic robot-like sweepers, and clothes cleaning will be electric, too.

■ Climate control will be complete. Humidity and temperature will be completely at the homeowner's command. Domed areas in the yard and swimming pool, as well as other devices, will be used to fuse the indoors and the outdoors in ways to suit each individual's taste.

■ New houses will have easily replaceable parts. The construction industry will have moved so far toward unitized, prefabricated construction that it will be a simple matter to redesign your house yourself when you get tired of it.

■ The kitchen will almost have disappeared. Small mobile cabinets, handsome enough for the living room, will house the basic cooking and refrigerating units.

Along with these new things, there will be the basic appliances: the range, the water heater, and the dryer. They may take different shapes—as refrigeration units today are far different from those of the early 1930s—but they will be needed and used. In addition, lighting levels will be raised, and more people will enjoy the benefits of better light.

People in the power business have a special responsibility to the future. The decisions they make today affect one of the basic backbones of the economy and have a direct relation to the kind of world people will have in the year 2000. Increased electric energy use and increased machine use are so closely interconnected that it is impossible to separate them. They are the Siamese twins of productivity. As energy use goes up, meaning that machine use goes up, per capita income rises. Similarly, energy use and energy costs are closely connected. As the price goes down, people use more energy. As people use more energy, the price can go down even more. Thus, if the price of energy can be kept down, it has a direct effect on living standards.

Appendix

Statement of Principles on Coordination of Electric Power Supply

The electric utility industry is a vital, continually expanding and evolving organism. The use of electric energy in the United States has been doubling about every ten years. The technology of electric energy supply has had to match this demand. Only a relatively short time ago a 100-mw unit was considered a large generator; today it must be 450, 600, 800 or even 1,000 mw to be considered in that category. Not long ago 110 kv was high voltage. Today, equipment is in operation at 345 and 500 kv; 750 kv is not far away; and 1,000 kv already is being visualized. These developments permit the larger individual systems and coordinated areas to achieve both higher reliability and lower costs.

The regulatory agencies and the public in general must be

assured that the industry is fully aware of and is meeting its obligation to continue to provide maximum service reliability to its customers at reasonable costs.

The Edison Electric Institute believes that:

1. The proper interconnection and coordination of electric power systems on an area basis is the most effective way of providing reliable and economical service in this country. We believe that reliability and economy of service are not contradictory when power systems and their interconnections are adequately planned and coordinated.

2. In this country there is a wide diversity among areas in population and load densities. There are also wide differences in the geographic, economic, and social characteristics of the several regions of this large continent. In general, electric power system interconnection and coordination have developed on an area basis, with many inter-area ties. Accordingly, the composition of these areas and the interconnections and coordination within them should be continually reexamined from the standpoint of total load, load characteristics, geography, population density, and number of participating power suppliers.

3. Once the appropriate area for coordination is determined and agreed upon by the participating systems, forecasts should be made of the area electric energy requirements for a number of years in the future by each of the members of the coordinated system and for the area as a whole. Studies should be prepared to determine reliable and economical means of supplying the area in light of the anticipated growth, and the technological advances that can be anticipated. Such long-range plans will need periodic review and change to take account of the advancing technology and changes in economic conditions. Since population, load densities, and other important characteristics will vary from area to area, it is to be expected that planning and design requirements of the several coordinated systems will also vary. However, many of the principles of coordination are com-

mon, and coordinated areas can benefit by exchanging views and information.

4. Each coordinated area should establish appropriate committees concerned with the management, planning, and operation of the coordinated system.   Proper liaison between areas should also be established.

5. Each utility should assume the responsibility for assuring that its own system is so planned and operated that it will be a firm element in a strongly coordinated area.   This will offer the best assurance that interruption to service will be minimized and that the most economical service will be provided to the customers.

6. Proper area coordination strengthens the supply of power for the needs of national security.   It also allows greater flexibility in locating critical defense loads.

7. There should be a continuing research program by the industry to develop economic and reliable higher-voltage apparatus, so that as the need for higher-voltage transmission develops the necessary equipment will be available.

8. The utilization of the latest technology in planning, building, operating, and controlling electric power systems to obtain economical and reliable electric service is and should continue to be the constant goal of utility management. Furthermore, utility management should not only continue to encourage equipment manufacturers and others to pursue the development of this technology, but should also take an active part in initiating and carrying out the required research and development.

9. Utility systems should expand and strengthen mutual assistance plans for providing manpower and equipment to assist in the restoration of service whenever disasters such as severe storms, hurricanes, and floods cause service interruptions.

10. Coordinated electric power systems should equitably share the costs and benefits of coordination in accordance with prevailing local and regional considerations.

Conclusions

The principles of coordination cited in this report are based on the experience gained by many persons over the years in planning, building, interconnecting, and operating electric power systems. The application of these principles has in no small way been instrumental in providing America with strong and reliable electric power systems. Electric power companies are continually expanding their coordinated operations to increase the benefits of economy and reliability to their customers. In addition they will continue the aggressive promotion of all possible uses and benefits of electric energy as it is essentially in the best interest of the electric utility industry and, more importantly, of the country.

Index